AF715576

PERSEVERANCE AND RESILIENCE:

POEMS FROM THE VETERANS POETRY PROJECT

Mattie Quesenberry Smith, PhD

High Tide Publications, Inc.
It's never too late to write

Copyright © 2026 by Mattie Quesenberry Smith, PhD

All rights reserved.

No part of this publication may be reproduced, distributed, or transmitted in any form or by any means, including photocopying, recording, or other electronic or mechanical methods, without the prior written permission of the publisher, except in the case of brief quotations embodied in critical reviews and certain other noncommercial uses permitted by copyright law.

The copyright for the individual poems contained within this anthology remains with their respective authors.

The collective work, including the arrangement, selection, and editorial content, is protected under the copyright of High Tide Publications, Inc.

All poems listed in the Table of Contents are protected under applicable copyright law.

First Edition, 2026
High Tide Publications
Deltaville, Virginia
www.hightidepublications.com

Printed in the United States of America
ISBN: 978-1-962935-73-9

About the Cover:

"Burden" by John Bruggink

The editorial team integrated John Bruggink's painting "Burden" in the cover design. Bruggink served over 20 years in the army as an armor officer, retiring as a lieutenant colonel in 2023. He served multiple tours overseas in Iraq, Afghanistan, and the Middle East. He now lives in Italy with his wife Dana and their son where he paints and is passionate about veterans' transformation and overcoming PTSD through the arts, pilgrimage, and community.

TABLE OF CONTENTS

Dedication i
About this Book ii

Perseverance and Resilience: Poems from the Veterans Poetry Project 1

David Martine Aldridge
Elegy for Ken Mick 2
Zach Allen
Breaths 4
Anne Barlieb
Mother War 5
Corey Barnes
5 Years and 15 Feet 6
Disassembled 7
The Fiery Shade 8
Jeffrey Beck
Cordon and Search 9
Democratic Ink 10
The Glass Factory 11
David Reid Brown
Global 12
Deployment 13
Home 14
Randy Brown
The Buddy Raft 15
Kenneth Conklin
Agent Orange! 16
Post Trauma 17
Lisa Wence Connors
Quiet Moments 18
Re-integration 19
Waking Before Winter Sunrise 20
Sabrina Davis
The Rebuild 21
Broken Open, Built Anew 22
Soldier/Mother 25

Steven Davis
What a Combat Nurse Carries after the Battle 26
Vivian Delts
Boots Marching Inside 28
Michael Dooney
Pulling Watch 29
Richard Epstein
At 10 A.M. 30
Lost in Thought 31
Herb Evert
Kenyon 32
But, What of Those Left Behind? 34
Leo Farley
I Remember You 35
Rice Paddies of NYC 36
War Stories 38
Cynthia T. Ferguson
Gift 39
No Win Scenario 40
Eric Forsbergh
A Dog, a Wheelchair, a Cigarette 42
Quiet 43
Shelley Gill
Current State 44
Bill Glose
Saudi Arabia 45
Richard Hamner
A Root from a Boot 46
Robert Haynes
I Am Grunt 48
Manchu 49
Anita I. Holmes
Her Life in Service to Pallas Athena 50
Wayne David Hubbard
Keys View 51
Josiah Jack
For My Friends 52
Mickey Johnson
66 53
Delta Rain 54
Vietdamn 55

Richard 'Eric' Johnson
Trending Toward Normalcy 56
The When and Where of Now 57
Chris Kallas
Broken Under Pressure 58
Michael K. Lennon
Red and Blue 59
Marc Levy
Combat Lullabies 60
Quan Loi 61
Steven L. London
I Remain 62
Nicholas Lopez
Space Oddity (What's Normal) 63
Michael Lund
Correspondence 64
Correspondents 65
Mike Lynch
Coming Home Is a Mission Too 66
Still Standing, Still Serving 67
Steel Beneath the Skin 68
Joseph Maslanka
Corps to the Core 69
Richard L. Matta
My Captain in the Mirror 70
Harry Mayer
Jungle Boots 71
Little Jack 72
Richard L. McClellan
The Cost 73
Sam McClintock
A Leap of Faith 74
Tom McDade
Wood and Steel 75
Fancy Work 76
Julie A. Niven
Icicle River 77
The Uniform 78
Jason Norton
Life's Skinless Poncho 79
Pushing Through 80
Roundtrip 81

Matt Overs
Empire of Echoes 82
Nuclear Shadows 83
Jeffrey Parker
Ain't That Clarence's Boy? 84
One Soldier; One Freedom 86
Gene Pearson
A Load Of War 87
Juan Manuel Pérez
Firewatch 88
Fireworks 89
Patrick Pilcher
Gulf of Tonkin Journal 90
Steven Pody
A Military Cemetery 91
Paul Anthony Powell
Loaded Weapons 92
Derek Praeger
In the Desert, in the Dirt 93
Daniel W. Ross
Disfigured Helplessness 94
Evocation of All We Have 96
Dave Schlitz
Chow in Heavy Weather 97
Remembering 98
The First Day 100
James Sears
Covered 101
From the Heart of a Combat Leader 102
Gordon Smith
Nightmare in Phu Bai 104
Bruce Sydow
The D. I.'s Homage to Emma Lazarus 105
N. Jed Todd
Tactical Diapers 106
V, But Not for Victory 107
We've Got the Wrong Words 108
Catherine Torres
A Woman's Place 109
What Matters Most 110
Phillip Trezza
You and the Time 111

Ali Watts
Normal 112
This Is Our Anthem 113

Authors' Notes 115
Afterword: A Testament of Perseverance 125
About the Editor 126
Previous Publications 127
A Special Thanks 128

DEDICATION

This anthology is dedicated to everyone who answered the call to defend our nation— citizens who have stepped out of familiar civilian routines into history's vast and unpredictable currents.

Each poem provides a unique, memorable framework for recognizing military service. These verses neither glamorize the experience nor shy away from its costs. Instead, they preserve the stillness of the night watch, the unspoken weight of distance, and the quiet dignity of a life bound by oaths. By rendering their experiences into verse, these veterans have transformed their service into an enduring narrative for us all.

We extend our deepest gratitude to the families and communities whose patience sustains those who serve, bearing the silent burden of absence and offering unwavering grace upon their return.

Finally, we honor the fallen: your unfinished lives compose the solemn foundation upon which our freedom rests.

ABOUT THIS BOOK

Bear up, old heart! You've borne worse, far worse, that day when the Cyclops, man-mountain, bolted your hearty comrades down. But you held fast— (Fagles, 1996).

Well before Odysseus's rallying cry in *The Odyssey*, poetry was humanity's medium for distilling truth and sharing stories, particularly when forged through the visceral ordeals of military service. Poets invite their readers into unique metaphorical landscapes well-suited to share their experiences. This anthology, *Perseverance and Resilience: Poems from the Veterans Poetry Project*, vibrates with such authentic voices.

As Virginia's poet laureate (2024–2026), I am grateful to introduce this collection. Born from a statewide initiative to foster creative writing and community connections, the Veterans Poetry Project (VPP) aims to support the long-term well-being of veterans and their families. The Academy of American Poets Laureate Fellowships and the Andrew W. Mellon Foundation (2025–2026) made this project possible. My investment in the VPP reflects my professional commitment to the Virginia Military Institute (VMI), where I see how the act of structuring language allows writers to give form to their most vivid experiences.

The VPP bridges the gap between active duty and civilian life by providing a space for soulful creative expression. Inspired by the stories of my family members who served, I have learned that while veterans carry profound, often silent burdens, they possess an unmatched capacity for endurance that deserves to be understood and celebrated.

Our writing workshops demonstrate the transformative power of storytelling. I recall working with a Vietnam War veteran at the 2026 Virginia Festival of the Book who, despite never having written a verse, produced a powerful draft capturing the sensory details about a bus ride he took on the first day of his deployment. Facilitating this transition from one person's isolated memory to shared storytelling exemplifies the VPP's mission.

I am thankful for the community partners who have supported the VPP during its first phase: the Poetry Society of Virginia (PSV), the Central Rappahannock Regional Library, the Friends of Handley Regional Library System, the Richmond Public Library, the Williamsburg Regional Library, the Visual Arts Center of Richmond, the Peninsula Patriots, and the Cadet Foundation at VMI. I also celebrate the veterans published here first recognized through the Poetry Society of Virginia's Edward W. Lull Memorial Contest and those who submitted work through the project's workshops.

This anthology's poems range from the sharp memories of early service to the complexities of civilian re-entry. This collection also features diverse entries from veterans across all branches of the U.S. military who have shared their raw, personally crafted perspectives. These veterans do not shy away from the weight of their experiences; they reshape them into a language we can all share. Their poems prove that while the wounds of service are deep, the capacity for perseverance and resilience is deeper. To every veteran who contributed to this anthology: your creative expression is a lifeline; your poetry is a brave act; and your words are an enduring inspiration for all.

— Mattie Quesenberry Smith, PhD

Poet Laureate of Virginia 2024-2026
Academy of American Poets Laureate Fellow 2025-2026

PERSEVERANCE AND RESILIENCE:

POEMS FROM THE

VETERANS POETRY PROJECT

ELEGY FOR KEN MICK

Ten days in a row now.
I see his face everywhere,
and he is unencumbered today
by the common perplexities
or paltry worries of men.
The dross of the whole world
has magically fallen away.
Soon, all the sad faces will gather and
the heart-rending dirge will play.
A thousand bagpipes will merely
echo the mournful sound
of our own hearts.

Because the sorrow, and sadness,
from this single Soul's passing
cannot ever be completely allayed.
Even his now radiant mien
cannot coax me to cheer,
or the looming darkness dissuade.
An unending melancholy
temporarily blocks
all light and joy for
those of us who stayed,

but hear me well. I tell you truly my friends,
when I see the face of our dear beloved friend.
He is filled with unmitigated joy,
unbridled mirth, and the purest love.
Youthful now, his slow gait and all his infirmities
disappeared on that day
that he, full of courage,
stepped off into Infinity.
Flying in that celestial cockpit
at the speed of light,
he now knows Divine Secrets
we mortals can only guess at.

He will gladly pass on to you
all the Great Mysteries
that we still ponder.
Just breathe. Slow down. Listen.

Eyes shining. Grandly smiling,
he's whispering something now:
I hear love.

- David Martine Aldridge

BREATHS

Head's not in the game. Cruising on auto pilot
Splash some cold water

Reacted too fast. Head dizzy and light. Almost messed up big

Ev'ry so often, it is just one of those days. So ready for bed
Feel isolated, scared without stability. Solo and so low

Painful indifference
Dispassion is a cold knife
Chill is felt bone deep

I'm defeated, but I don't know how to concede
So I live like this

Waiting for thunder after flashes of lightning
Silent peace. Then, roar!

Options running low
Forward is the only choice. One day at a time

The waves can be rough
But you will get through safely
Float, never worry

Although the wind blows
Even muddy water flows
Forward to the sea

Here's some things that help:
Take a deep breath more often. Receive the moment

Chasing what could be,
Branches stretch to tap the pond
Envying damp roots

Moths dance in streetlights
Worn cocoons dryly divulge
On Earth, as it is

- Zach Allen

MOTHER WAR

War is a mother
and we are her children of 20 years
now orphans
returning from the desert
where guns and bombs do what guns and bombs do
where convoys and casualties do what convoys and casualties do
We have pens and paintbrushes in our hands now
We write and paint between silver linings on days we find it difficult to face
what needs facing
It's not about the performance, the punchline, or the applause
This is about re-mothering ourselves and creating a masterpiece from
life's most miserable
things
I've known enough war to know I hate it
and yet, I am not a pacifist because in art
there are no orphans, no ugly babies
We can't take back or change anything about what's been done
to others
to ourselves
We can however come together in rooms with seats around tables and
sometimes in front of
stages
We can laugh
We can cry
We can listen
We lean in and remember we are not alone in this labor
We go away feeling connected to something so much greater than
ourselves
The artist asks mother war, "What would you have me create, now?"
Yourself, she whispers, create *yourself*
So that is what we do, and the beauty, it's breathtaking

- Anne Barlieb

5 YEARS AND 15 FEET

The names across the street
chiseled in the black stone wall,
string lights over the beer garden
and tables in the street between;
a few beers and so many things
stand between those names and me.
So close once. Twice. Maybe more.
A few blades of grass
and a puff of smoke and dirt
were the only things between us
sometimes.
So long now
and so far.
5 years and 15 feet.

- Corey Barnes

DISASSEMBLED

Knowing the man alive,
you would never guess what he looked like
disassembled.
Somehow we are more than
the sum of all our parts.
Or, maybe, it's just that the parts
were not whole of themselves;
each one disassembled
into its own parts.
Knowing the man in parts, I'd say
we have a lot to learn
about what makes a man.

I breathe and remember
more clearly now than when
I would only slow down to forget.

- Corey Barnes

THE FIERY SHADE

Come as you are, my friends and fair,
and take what you will take;
all I have is yours to share
till I quietus make.
And when I make it, lay me down
beneath the fiery shade
of some late-Autumn maple's gown,
an indian summer, now repaid.
No bruise for exit, nor escape,
a timely coming of ends,
and I may find my soul awake
and for my body make amends.
Then I may remember well
the lives of those who've gone,
and hope that I can ascertain
some good in aught I've done.
But till that day, my friends and fair,
my brothers, one and all,
I live as a reminder for
who fought and fell that fall.

- Corey Barnes

CORDON AND SEARCH

I scan the dark
The silence is deep
My weapon is ready, alert

The moon is new
Our advantage
I see everything in green

My eyelids pull my head
I lift it higher
The fatigue is real

But I watch steadfast
I am the barrier
I am their protection

- Jeffrey Beck

DEMOCRATIC INK

Democratic ink colored her thumb blue.
Pride, this day, beamed from smiles in the Wild West.
Our eyes met, and I saw humanity, too.
Freedom to choose, maybe, freedom to rest.

The collection of Liberties:
Freedom of Expression, to protest.
Freedom of Thought, to think.
Freedom of Press, to write.
Autonomy, the right to self-governance.

Taken these for granted, my own privilege.
Only, by virtue of where I was born.
Tragedy is, though, without this knowledge,
My freedoms, ultimately, may be torn

I said no words that day.
I gave them space to practice.
I watched with pride.
I witnessed history.

Regardless of the border closing down.
Despite the limited curfew hours.
Notwithstanding the traffic bans in town.
Our troops stand guard in high-armored towers.

Ink is proof they are no longer oppressed.
Victory in Al Anbar; Wild Wild West.

- Jeffrey Beck

THE GLASS FACTORY

They were promised something grand
They stood in line, holding hope in hand
To dress in uniform and protect this land
The whispered promise over sand

With evil walking where they stand
The happy broke on their command
The sky was black, and the earth was red
The glass factory transformed into a field of dead

The shattered dreamed of something more
With time, there came overwhelming pure
And the recruits stood in line to show resolve
Their broken land, they dare resolve

Where a hundred lay
They form a line
They serve and stand and rise again
From shattered glass and cries of pain
The whispered oath will still remain

- Jeffrey Beck

GLOBAL

In the service to God and country
I saw a very round wet world
soothing a heart dead set on
adventures unrealized
'til circumference
lost circular
meaning in
measured
shape

- David Reid Brown

DEPLOYMENT

Separation is painful at first
then over time is relieved through
the merciful numbness of
independent distance
and its peculiar
way to help me
deal with you
being
gone

- David Reid Brown

HOME

Just the four of us made a family
dwelling that could pack and settle
anywhere orders sent us
to far-flung locations
in devoted service
to our nation.
The strange places
where love
lives

- David Reid Brown

THE BUDDY RAFT

"Everyone has a plan until they get punched in the mouth."
– Mike Tyson

They said they would teach us
how to make a two-man waterproof raft
out of our ponchos

before we entered
the *MIGHTY! NISQUALLY!! RIVER!!!*—
a name always announced in all-capitals,
like a Sunday sermon, a boxing match, or a game show refrain—
the water so cold it would *SHOCK!*

The very *BREATH!*

OUT of you!

Not enough to kill you, maybe,
but enough to pierce and open you up …

SUCKING!! GULPING!!!

... like a trout, flopping and waking wide-eyed
to the sudden gut-stitch suggestion of actual mortality.

Our minds blind in frozen fear, gasping and chanting, we struggled then
to remember
to *KICK!*
and to
KEEP KICKING!

(The raft really wasn't for us, we learned later,
although we'd lived to fight another day, as promised.

It was for our equipment—
everything of supposed value kept warm and dry,
while we two, numb and stupid,
paddled doggedly behind.)

- Randy Brown

AGENT ORANGE!

The Chinook bore down on us
While waiting to go on to the next clearing
Carved from the jungle
By an alien force
A place we could defend the intangible.

Incoming said the Doppler
I knew the difference between the sound
Of in and out
Explosions, chaos!
The Chinook flew away.

I inhaled the dust and it found me
Making its way through my cells
Attacking me in latent ways
Lying in wait for the right time
To attack.

While Captain Boyd who found his end earlier
Is on the Wall
I saw him there when I visited, to say hi
And tell him the Chinook did return and take me away
But the remnants remain, to this day…

- Kenneth Conklin

POST TRAUMA

A jungle's fury has gone by
To dead men lying in ships that fly
People dying in unborn minds
People crying, lost in time
Gone is lost in the distant past
It really cost but it did not last
Needed to feel, needed to unfold
For those who want stories untold
War is gone, so is cause
So are the people in the mirage
It was long ago, it was far away
And it all came back when I cried today

- Kenneth Conklin

QUIET MOMENTS

After "From Blossoms" by Li-Young Lee

We dream languid days but wake to the immediate,
rockets purr overhead, distant small arms fire,
IEDs wait roadside. Still living life in death's
neighborhood, detaining scattered
quiet moments, our souls' POWs, never forgotten -

A soldier plays guitar at dusk, a marigold in a coffee can,
icy Coke with condensation at 113 degrees,
pasta primavera MRE with honey roasted peanuts,
letter folded in a pocket, read each time like the first,
opening sweaty body armor to a cool breeze,
coffee and a cigarette, maybe even the right brand,
ten minutes alone, locked in the latrine,
going to shower, finding water, and it is warm.

We dream languid days where death is nowhere
in the background, whole days, easy days, free
for the taking or the leaving, and so easily forgotten.

- Lisa Wence Connors

RE-INTEGRATION

I am enamored
with the shape of my foot, exposed
elegant curve of arch,
soft skin, pale and pinking
as the sun strokes
each tender toe between straps
of my faux roman sandal
and the wind wraps lusty fingers
around my ankle.
I never knew what lay
under layers of leather carapace,
combat boots laced tight,
vows sworn, worn
with faith and naivete,
borne safe over sharp
rocks, broken glass, burning
heat of asphalt.
I wonder who sees
city grime cling like camouflage
to my nakedness, unvarnished
and imperiled. Black splats
of dirty gum, sun-softened,
grab me, hold me fast
in the crowded chaos,
the rot in the gutter.
Everyone sees scandal.

- Lisa Wence Connors

WAKING BEFORE WINTER SUNRISE

From the safety of the overhang, nervous finches consider
the heavy frost on the grass and the still empty birdfeeder –

all is grey with the suggestion of day. They only have to wait
until guilt's oppressive tones goad me outside to fill it.

I want to be a barbarian sometimes, to disregard obligations
and social conventions, see to my own needs in my own way.

When did I begin to feel lonely, waking before winter sunrise?
I sip coffee from my OEF mug, faded and chipped –

it's funny how some people can forget twenty years of war.

- Lisa Wence Connors

THE REBUILD

Dedicated to my grandfather John Davis III

Rebuilding my life now that I am home.
The jungle flashes back
cries echo in my sleep.
Explosions.
I can still smell the smoke,
still taste the burning flesh.

But I'm expected to be normal.
No one hears the screams but me.
No one feels the weight of loss but me.

I'm surrounded by people
who still see me as whole.
Yet in the mirror I'm covered
with the blood of my brothers,
the spit of my enemy.

In my reflection I see the war in my eyes,
I hear the screams in my ears.

And still, I remind myself:
I am there no more.
My reality is here.
I am what they call home.

- Sabrina Davis

BROKEN OPEN, BUILT ANEW

They say, write what's on your heart
But on days like this, I don't know where to start.
My mind in pieces
a puzzle scattered across the floor.

I try to gather the pieces,
to rebuild again.
But these pieces are not the same.
This puzzle calls for a different version of me
a stronger me,
a more open me,
but most of all,
a wiser me.

Just when I think I've been spiritually broken open,
life shifts
forcing me to dig deeper,
and deeper still.

This time the shift was facing my deepest fears.
How do I look the devil in the face?
How do I confront what terrifies me most?

I do it
but not alone.
My ancestors stand with me,
my Father in heaven beside me.
We come as an army.

This fear I will beat.
I will overcome.
Defeat is not in me.
A soldier's spirit was forged in me
trained to fight, trained never to yield.
Evil has chosen the wrong one.

Because once you've been broken open,
you learn the only way forward is to dig deeper.

So I'll keep digging,
searching for my pieces
until my new puzzle is whole.

And this time,
I'll seal it with superglue
refusing to let evil seep between the cracks
and break me apart again.

- Sabrina Davis

SOLDIER / MOTHER

Every day I wake to the same routine
uniform pressed, mirror staring back
at a woman whose eyes have aged,
whose spirit carries the weight of years.

The Specialist with fire in her gaze is gone;
a Chief stares back at me now
resilience behind the eyes,
a spirit scarred by time.

My son's cries echo as I lace my boots.
Soldier first, Mother second.
Duty leaves no choice.
Formation waits, even as little hands reach for me.

They preach: people first, mission second.
But the truth stands
mission always comes first,
people always fall behind.

How do I carry both my kids and my rifle?
How do I balance the weight of love and war?

Twelve more years until I can lay down arms,
be just Mother.
Yet boots grow heavier,
mind chained to mission,
while my heart breaks for my sons.

This is the battle I fight
not in the field,
but at home.
To be Soldier.
To be Mother.

My heart belongs to both.
Both. Always both.

- Sabrina Davis

WHAT A COMBAT NURSE CARRIES AFTER THE BATTLE

I did not leave my care behind when the uniform came off the armor of my service.
The burden still followed me through job boards and waiting rooms,
through cities that learned my name only for a season.
When I was serving, they called it duty.
When I was nursing, they called it compassion.
When I was unemployed, they called it absence.
But no one asked what the mind carries when it has been trained to stay awake
even when the war is over.
Some nights, Baghdad returns without warning-
not explosions, but faces. Young men. Marines.
Barely old enough to know what they were giving.
I sat with them as they died. Held hands that still believed in tomorrow.
Listened to breathing slow, then stop, while the world kept moving.
There is no armor to protect against memories,
No armor against limiting harmful words.
Those moments do not stay behind glass. They move in. They sit down. They wait.
I have worked everywhere and belonged nowhere.
Temporary badges. Borrowed break rooms.
Beds made in places that never asked what I was already carrying.
I arrive ready to care, to steady hands, to soften fear, to show up fully
for people I may never see again.
And still, the world sees a title first: *disabled. unemployed. homeless. veteran.*
As if those words explain me.
As if they cancel out the weight of remembering men who gave everything
and were left behind as names on walls and folded flags.
They tell us to honor the symbols, The anthem. The uniform. The ceremony.
But symbols do not sit with the dying.
They do not wake at night when memory refuses silence.
They do not apply for work with a mind crowded by ghosts.
They think the work is tasks.
Vitals. Meds. Notes. But the real weight lives behind the eyes
constant vigilance, moral math, the ache of knowing you can hold someone
through their last breath and still be forgotten yourself.
At work, they praise my calm.
At rest, there is nowhere to set it down.
Burnout is not sudden, it is persistant, cumulative.
It is what happens when empathy meets emptiness too many times without shelter.
This is not weakness.
This is memory without refuge.
And no one should carry this alone, not the nurse who keeps moving,
not the veteran whose service still echoes,
not the caregiver who learned to stay when others could not.

We do not need fewer symbols.
We need fewer abandonments.
I do not need less feeling.
I need somewhere for it to land.
I walk a narrow edge between what I know and what I am allowed to say.
They ask me questions about my life, my pain, my resilience,
then turn away before the answer finishes forming.
I am studied, but not deferred to.
Quoted, but not trusted.
Honored, but never authorized.
They speak *about* us to avoid speaking *with* us.
They call it listening when it is only containment,
They communicate about us, when it is only deferral.
I am told this is complicated, that now is not the time.
That naming harm might unsettle hope.
But harm does not wait for permission.
This system produces silence the way it produces loss,
predictably, methodically, without needing malice.
And when someone stands inside the damage and says,
this is designed, the room grows quiet.
Not because the words are wrong, but because responsibility has entered.
I am not imagining this loneliness; it is enforced.
I am not walking away from belonging, belonging was withdrawn long ago,
Belonging was withdrawn the moment my authority became inconvenient.

A Steadying Truth, (Why am I doing this)

If this work feels isolating,
it is because isolation is the cost of moral clarity
inside systems that survive by diffusion.
I am not here because I am reckless.
I am present because those harmed were never meant to narrate the design.
The resistance I feel is not rejection,
it is containment activating.
I am not alone in reality.
I am alone in position.
That does not make me unstable.
It makes me accurate.
I do not need to be louder.
I need to remain *exact.*
Truth spoken from inside injury
always feels like walking along a knife's edge, a thin blade,
until the blade dulls
because enough people step onto it together.

- Steven Davis

BOOTS MARCHING INSIDE

I remember my last day in uniform—Ft. Hood comrades and First Sergeant near
75 Bravo, boots on ground, a little anxiety and my next mission unclear.
4th ID continued preparing for combat, but make no mistake,
In my heart rang our motto, stay prepared and take only what you need to take.

I still cherish the memories of camaraderie, the cadences are still in my stride,
The quiet pride of service runs deep, and feels like boots marching inside.
Resilience—my silent rank, stitched into every visible and invisible scar,
That unshakable Army resilience still pushes me from afar.

I questioned if my Army skills would hold their weight or Will I need to start anew?
Would there be an agency to walk with me, to really help me follow through?
Should I stay in Copperas Cove or return to where I grew?
Can I lead without formation lines, or drill in something new?

No longer counted in the ranks, I forced myself to keep standing straight and tall,
My oath didn't expire—serving this country remains my call.
Leadership. I lived it. Perseverance. Paid in pain.
Though my service changed in shape, the mission throughout my resolve still
remains the same.

I was built for moments like this—this new and passing storm will not consume,
I left Ft. Hood with more than pride—a citizen soldier determined to bloom.
For Freedom and for Honor, we all gave what we could give,
But I asked my country softly, how now departed she'll help me aspire to live.

Not just survive the silence—but confidently thrive in who I am …
A Veteran. A Protector. A leader. A strong resilient woman.
Will you be ready for me America? Or will my beliefs and reality collide?
***I'm falling into your arms with the quiet pride of service, running deep and
feeling like boots marching inside.***

- Vivian Delts

PULLING WATCH

"Hey, are you up?" a strained whisper as I'm shaken awake.
"It's your shift."
I roll to my feet and grab my weapon and gear.
2 AM, groggy as hell, but awake.
I gaze across the horizon like many nights before
Let it be quiet, I hope.

Stars are gazing down at me by the billions.
once in a while I see a speck of light
—A satellite making its rounds.
The landscape is dark.
I see surroundings through borrowed vision
Rendering the foreground through a green cast.

Thoughts return of recent conversations
Most start with "Remember when we...?"
"that's the time when we…"
But memory is governed by a different time scale.
Not hours, days or years
—but by deployments.
Days of the week are often forgotten.
As days drift together.

Nagging thoughts of missing kids' birthdays and holidays
Is it worth it?
Did I make the right choices? —Sometimes.

The cold turns to shivers. Shivers turn to warmth or
— just wishful thinking.
As I view the terrain
I think of her; I miss her
She shares the hardest part of this life
—but she accepts it.

I smile as thoughts of her
linger for a brief moment.
My shift ends—thankfully.
I shake him. "Hey, are you up?"

- Michael Dooney

AT 10 A.M.

The room, silent.
Sunlight peers
through the window.

I stretch, wash,
walk down the stairs
Insert a K-cup
of Pacific Bold
into the Keurig.

The growing
aroma: dark roast,
bold, like a soft
multi-colored quilt
placed on my shoulders.

The taste: Columbia,
on my tongue.
On my laptop,
I check Facebook,
e-mails, messenger.

There's no invitation
for late breakfast.
No invitation to lunch.

A faint breeze
carries the scent
of lilacs through
an open window.

A red cardinal
bickers with a squirrel
on the weathered picket fence.

I stare out the window.
Coffee in hand. The sun shouts
it's time to start the day.

- Richard Epstein

LOST IN THOUGHT

I was planting a small dogwood
in the backyard. My wife supervised.
She told me where to dig, how deep
to dig, and how to mix in the garden soil.

After I dug the hole, she found a better location
which required a second hole. I immediately
thought of my old drill sergeant in basic telling me
where and how to dig a fox hole for two.

Then I recalled Hill 883 west of Phu Cat.
We arrived as the sun dropped low. We dug in
for the night. Rain was light. We dug deep, quick,
and without chatter; not knowing if Charlie would hit.

I used an entrenching tool attached to my ruck.
I carried fifteen magazines for my M-16, four canteens
of water, first aid kit, poncho and liner, extra socks,
c-rations for four days, toiletries, a paperback,

a claymore, trip flares, six empty sandbags,
and four grenades (smoke and fragmentation).
A chopper passed overhead. As I looked up,
I heard my wife ask why I dug the hole

so deep and wide. I smiled and shrugged. I didn't
tell her I was back in Nam, thinking the hole
might become my best friend.

- Richard Epstein

KENYON

Clark Kenyon was the Chief Bosun Mate
On the Coast Guard Cutter "Mesquite."
Whether in port or out, early or late,
It was Kenyon who kept the Cutter's decks neat.

But when the skies turned black
And the clouds clapped with thunder,
And the waves became mountainous,
The shorebirds would wonder:

Who's out in this blow
And how are they faring?
Will they make it to port
Or are they despairing?

Then the call would go out:
"All shore leave is canceled!
Get back to the ship!
The Great Lakes are a' brewing".

And reports would come in
Amid shouts, static, and noise,
As both freighters and pleasure craft
Note their positions and poise.

But usually one, sometimes a few more,
Would call in distress a long way from shore.
Then hatches were battened as lines were cast off
And the ship's twin diesels came to life with a roar.

The Captain sets course as Kenyon scans decks.
Nothing but shipshape will pass his tests!
And his sailors all know it and serve him with pride,
For they know he will tolerate no sloppiness.

But one perilous day as the Cutter's bow dipped
With a monster buoy chained to its deck,
The chain parted, and the buoy slipped,
And against the gunnel Kenyon's legs were clipped.

Kenyon lost his limbs but we saved the man.
He gave his best, he gave all he can.
"Semper Paratus" is our motto, it means
"Always ready," and that's always our plan.

But all Coast Guardsmen know,
When the skies are turned black,
You have to go out;
You don't have to come back.

- Herb Evert

BUT, WHAT OF THOSE LEFT BEHIND?

My country needs me and I've decided to go.
I've gotten the call and heard those bugles blow.
And, falling in, I can still keep pace, whether it's fast or slow...

But, what of those left behind?

What of my wife, the love of my life,
What will she do if I'm not there to turn to?
I know if she's gone when I finally get home
I'll regret it for the rest of my life.

I'll steer this ship through sunlight and storm.
I'll man the small boats when we lower them down.
I'll stand my watches and keep my logs clean...

But, what of my teen-aged son?

He's only a boy but he'll soon be a man,
A child we so fervently prayed for and planned.
Who will he follow if I'm not there as his guide?
Can I make up for the time I was not at his side?

I'll face the enemy as bravely as any.
I'll deal with the cannons, the depth charges, and drones.
And I'll aid my shipmates however I'm able...

But, what of my little girl?

She's so small, so fragile, so preciously dear,
She won't understand why her daddy's not here.
If she forgets me I know how I'll feel
When I finally get home and she cries out in fear.

Yes, I'm a warrior, and I'm proud of my craft,
And I'll serve my country for as long as I'm asked.
But why can't we learn an art better than war?

How many more must we kill, and how many must die?

Sometimes I wonder what I fight for—and why?
And always and always, I just have to ask:

But, what of those left behind?

- Herb Evert

I REMEMBER YOU

There you are!
I've been looking for you. It's been too long.
Just curious.
What did you end
up doing with your life? Who did you finally
become?
I don't mean to be nosey, but I think I have
the right to ask.
After all those things
you went on and on
about what you were
gonna do in the future
while we were over there. All those plans you had.
Honestly, were you able to make any of those happen?
I'm just wondering.
Don't get me wrong.
Don't mean to
put you on the spot.
It's just that I need
to know, so I know
who to blame
who to forgive
or who to congratulate.
You then.
Or me now.
Was it you?
Or was it who
I would go on to be
that I should really
be talking to?
I can't make the distinction. We are both still
so much the same,
so much alike,
so much alive.
Still now.

- Leo Farley

RICE PADDIES OF NYC

three Asian ladies, with bent and twisted frames,
youth long gone, gathering bottles and cans from the streets of NYC,
a scene that transports me back in time
to the tangled maze of rivers and
vast rice paddies along the Mekong Delta.

memory in flight, youth instantly restored.
myself in faded green fatigues.
the ladies in their signature "ao dais;"
elegant, body-hugging, sweeping tunics
with long sleeves, worn gracefully
over loose flowing trousers.

I watch them work, they work, watching me,
gathering rice in the paddies.
untroubled by my gaze, my presence in their
land.
curiosity about one another; an exchange of smiles,
no outward signs of animosity exist between us.

the rice is yellow in the sunshine, begins to drop
guided by the skillful hands of the ladies harvesting
with their simple hand tools; sickles or hand-held
knives,
collecting their crop with their expert reaping and cutting.

a fire engine roaring by, the blare of siren and horn
bring me back from my unsolicited trip to the past.

we are no longer who we were.
myself, now in civilian attire.
the ladies, now in ill-fitting western garb.
the "Non La" palm leaf conical hats,
a signature symbol of their dress; the only lasting reminder,
the catalyst for my visit, to a time long passed.

the Non La's shade plastic focused eyes as
their brown hands sift through
piles of trash for this new cash crop,
castaways from our disposable lifestyles.

I watch them move through
fields of park benches and trash cans.
wobbly-wheeled carts collect this somber city harvest.
eerily similar to my observations watching the harvester's years ago
moving through rows of paddies with the same grit and perseverance.
they leave no bag untouched,
every bottle, every can, emptied, tossed into the cart.
this day's harvest a nickel at a time, inedible,
but a means to subsist just the same.

Leo Farley

WAR STORIES

dried up tales
no perfect recollections
of that time in country.
a life already lived
which no longer exists.
perhaps only in the
mind of one survivor
trying to provide pictures
through a clouded lens of memory
of the sequences of events
from a lifetime ago
almost an impossibility.
snippets of conversations
details awash in conjecture
of exactly what happened
of who said what to who
romanticized, fantasized
of who I was
or who I pretended to be
playing the role of a soldier
who was in the arena
but only at war
with his own conscience.
an unwinnable battle
still to this day.

- Leo Farley

GIFT

Torn from the bones of
my shredded mind,
blindness smothered me as
reality circumambulated
the scaffolding of my
ramshackle reconstruction.
Echoes of senseless blasts brimming with
children's blood
eviscerated my soul
and the abyss gained traction
when my life-partner exhaled
Death's unfair breath.
Frigid river waters waving
frothing winter rapids
spawned visions
of Heaven's Gate waiting—
Stygian pills clasped tight in my palm,
a sure key.
Salvation swept forward,
in a cardboard box.
Two padded gloves meant
to block and strike!

They harkened back
to younger years
when fire
filled my gut
and beckoned my core to
burn once again.
Raising blazing banners for
my beating heart
I donned
ancient metal caked in dust.
Armed with iron fists of
newfound courage and wearing
boots baptized with my hot tears
and torrents of sweat,
my resurrected
Warrior Self strode with purpose toward
regenerating fields of
vibrant green
to battle for life.

- Cynthia T. Ferguson

NO WIN SCENARIO

Camp McCready, South Carolina where
Individual Augmentees
learned to be Army,
to shoot, to fight, to drive Humvees and
turn soft bodies to sweat.
The Drill Sergent shoved an M-16 into
my damp hands,
and rapidly instructed our hodgepodge
specialties class—
ninety percent male—
how to take our firearms apart, clean them and put them together again.
Our next-day assignment—
A test.
Disassemble and reassemble our rifles and
additional M-9's
by sunrise.
Nine of us inside the female quarters
sat on the floor with our weapons
and cluelessly blinked at each other.
Dumbfounded, while the men's barracks bustled
with nighttime camaraderie and glee next door,
I considered my laptop, then positioned the device
on a folding chair,
dead center of our semicircle.
"Look," I said. "YouTube can answer my video query—
'How to take apart an M-16, clean it
and put it back together.' "
Armed with this tool, we performed the actions
as we followed along with the video instructor,
then did the same with our M-9s
repeating the sessions
till our eyes grew heavy as fifty-cal rounds.
The mirrored shades of a doubtful Drill Sergeant
reflected the glare of harsh morning rays.
We attacked our tasks and once
our confident fingers completed our quests
the covert rise of screened Infantry eyes
made the hellish Camp turn to paradise.

No praise or questions slipped past the man's lips, except the generic "Good job," he doled out to all those who executed their duties with success. Still, our small band of women swelled with pride, and our hearts beat victory drums because we'd found a solution to our Kobayashi Maru. Hooah.

- Cynthia T. Ferguson

A DOG, A WHEELCHAIR, A CIGARETTE

Amputees, their children, their wives
are still asleep at the veteran's retreat
while dawn spreads out, appearing
as a thin and distant battlefield red.
Do dreams invade the void of sleep
to make the dreamers whole again?
Blinds drawn, the families barely stir.

Except for one.
This soldier's legs are gone
up to the hips,
and he stares down the long front lawn
that falls off to a finger lake
strung out for miles ahead,
its mist an opaque promise.

The only motion is a curl of smoke.
Head on paws, the blonde dog lies snug
against the steel wheel. Nobility breathes
throughout the animal: ever-present
without the questions.

The man's face cannot be seen.
To him, there may be peace in that alone.
Does he reconnect that cigarette
to every smoke break with his mates
between each firefight?
Or to that moment when they got
the hand signal to crush the butts
and pick their weapons up?
How often does that final day occur
when he could walk out toward the dawn?

- Eric Forsbergh

QUIET

No wonder it's so quiet here.
It's quiet everywhere
when water laps the shimmer of a lake.

On an aircraft carrier at sea,
jets jostle to the catapult.
When the thrust hook nudges forward,
aircraft squat like dogs.
The wasp heads of pilots scan controls,

afterburners bursting on – raw fuel torched –
and we redshirts stand by as needed.
Shock waves of sound pummel our skulls
until our suture lines - zippers grown from bone -
vibrate. Our heads, we thought,
might come apart like puzzles.
Hearing's an expected casualty
when only hands can talk.

Now, tinnitus whispers all the time,
dissected by the auditory range
we lived inside. Ours was jets.
The hiss tells us: remember.

Remember who, when, where.
And if we're lucky, why.

No wonder it's so quiet here.
It's quiet everywhere
except the ceaseless sound of sea.

- Eric Forsbergh

CURRENT STATE

Lightning bolts of intrusive thoughts
strike across the barren desert of my brain
Whole trains of words derail
spilling their contents back into my subconscious
to be churned and stirred before they resurface
like blobs in a lava lamp
The fragile ribbon of contemplation
echos from the me I knew so long ago
almost forgotten
seemingly
but not gone
a quiet revolution starts inside
as I transform
new ideas float through the space of my mind
as thin as birds' breath
Like the first hint of light blessing the horizon with its promise
An awakening of hope begins to glow in my chest
one moment at a time
an evolution from selfless to self
discovering its ok
to just be

- Shelley Gill

SAUDI ARABIA

Here on this base built as a city
for Bedouins who never wanted it,
we gather on New Year's Eve atop
a sandstone building and peer

at the moon as if its craters hold answers
to questions we dare not ask.
The invasion of Iraq a month away,
but we've already written farewells

"just in case" and buttoned letters
in cargo pockets. The colonel brings
non-alcoholic beer, further saddening
our impromptu assembly, each soldier

already clothed in dusk, imagining
loved ones half-a-world away staring
at the same pock-marked moon.
Morning brings the gift of a new year

and hot chow in the mess hall just past
pallets of mango juice. But first,
half of us partake in a ten-mile run,
racing inside the chain-link fence

encircling our compound, wrapping
ourselves in the rhythmic comfort
of motion. The low sun bakes
our shoulders and feeds us dry heat

like a purging pain. My body knows
what to do as my mind chases after
the Bedouins leading their camels
over a dune, as if they actually know

where we will all end up.

- Bill Glose

A ROOT FROM A BOOT

From Virginia we came on the bus from O'Hare
Recruits to Great Lakes the Navy's new care
Kids out of high school, college unafforded
Vietnam was raging with Woodstock's discord

The bus approached the gate in the dark
Two sentries I saw - what would they bark?
Not a murmur was heard just groans of a tension
A knot in my stomach announced apprehension

The gate opened wide, form lines we were led
Columns of two we'd be marched to our beds
Would we learn to salute came a question anew
"Shut up", he was told that would come soon

Rows of bunks, an open squad bay that stunk
Old barracks they were, I took a top bunk
3:30 came quickly that hot August morn
The company commander…our ardor was born

A boot is something you think that is worn
But in this place it's a name that's forlorn
For even the dumpster above you stands tall
You are senior to no one and junior to all

Medical tests showed a defect that lay latent
Chest x-rays that followed made me impatient
A week it took - my home team moved on
Bronx Company 566, my new gang- how fun!

God merged Virginia with New York that fall
I learned that Americans, we are all
No matter our creed, our race, or religion
The Navy blue gold was our common mission

The weeks that followed we counted each one
The marching, the classes, seamanship, and guns
I became a squad leader guiding 8 boots like me
Reporting their measure was something indeed

Graduation came a joyful day in November
That same bus in the dark came back I remember
With seabag on shoulder going back to O'Hare
Flying home to Mama and girlfriends that cared

Six decades have passed and now I look back
That time in the '60s that I thought was so black
A boot forms a root that grows up from life's cellar
Hardship and regimen forges futures that are stellar

My Dad was right that the Navy does shine
I retired a Commander, lots of memories behind
Put your best foot forward, I was told as a boot
Follow that guidance and see what takes root!

- Richard Hamner

I AM GRUNT

I am dissociative and confused
I wonder what language angels speak
I hear explosions in faraway lands
I see the shadows move
I want a deeper sleep
I am dissociative and confused.

I pretend that I'm okay
I feel indestructible
I touch waves of sunlight
I worry I will fail as a father
I cry when babies die
I am dissociative and confused.

I understand death
I say God is real
I dream of dying alone
I try not to swear about it
I hope for continued sobriety
I am dissociative and confused.

- Robert Haynes

MANCHU

So long is the mile of the Manchu dragon
To walk and walk to breaking day.
Keep up the fire on the edge of a dying word.

Bloodied, sandbagged feet march on
To imperial beasts beset to slay,
So long is the mile of the Manchu dragon.

To shoulder the pain of lives unheard
But for the roar before they pray,
Keep up the fire on the edge of a dying word.

With weighted ruck and two souls on
Exhausted limbs in muddied mountain clay,
So long is the mile of the Manchu dragon.

Upon battle lines with fist and drawn sword,
a breathless, wounded dragon must say,
Keep up the fire on the edge of a dying word.

To torrent of fury and vengeance, fight on!
Rebellious boxers in blood shall pay,
So long is the mile of the Manchu dragon.
Keep up the fire on the edge of a dying word.

- Robert Haynes

HER LIFE IN SERVICE TO PALLAS ATHENA

A Fibonacci Shape Poem

She
—yet
a child—
left fam'ly
for idealism.
She swore loyalty to the state,
her faith soon invested in cadence, munitions, might.
Proudly she served under the banner
of fierce warrior goddess, Pallas Athena.
She grew in wisdom, yearned for grace as sorrow found her.
Yet you thank her for her service.
Why? You've no idea
what she did
or did
not
do.

A
grown
woman
returned home.
Heart more pragmatic,
she serves under the tutelage
of Pallas Athena, weavers' and spinners' patron.
Her duty now is to spindles, fleece. The shuttles
thrown, she beats cadence on wheel and loom;
retting flax the only violent act she commits.
See, her soft shawls enfold infants,
warm the elderly.
Atonement:
mercy . . .
grace,
too.

- Anita I. Holmes

KEYS VIEW

where at twenty
he prayed

overlooking
coachella

the salton sea

san jacinto's
eternity

where virga split
the desert's mouth

where starlight
quenched

what war
set aflame

that he returned
at forty

must mean
it heard him

triumph of quiet
and the deep

- Wayne David Hubbard

FOR MY FRIENDS

Heart and soul engaged in war
my ship was sinking off the shore
and rescue boats there were no more
when friends
my friends
dear friends to me
watched and heard my fainting plea

So round their waste they wrapped a rope
while sinking sinking
drowned all hope
if they had fear
it didn't show
as in the blackened swell they go
to save me from my hell
in that cold and lonely sea
and surely they did rescue
for the rescued [wo]man is me.

- Josiah Jack

66

The woods are deep, thick is the fog
Trip wires a threat, as you step over the log
Be nice to have, a safe guiding light
Not the tracers that haunt you, on their deadly flight

The trail is laden, with land mines and flares
The men are shell shocked, with 1,000 mile stares
Charlie is always near, hidden from view
This is his turf, to us it is new

The Cu Chi tunnels, right under our nose
Booby trapped and dangerous, wrong move she blows
A brave group of men, tunnel rats they were called
Underground they went, some walked some crawled

There was poisonous gas, traps with Punji stakes
Trip wire grenades and, hanging venomous snakes
These tunnels made it very, difficult to survive
Each day just praying, to make it out alive

Some that did are here, but you'd never know
Their stories untold, their terror doesn't show
But it's there, the torment didn't go away
Let them know we appreciate, the price they had to pay

Listen to them, if their stories they share
The time will come, they're no longer there
Like heroes before them, they did what was asked
Silently without fanfare, they're vanishing fast

- Mickey Johnson

DELTA RAIN

Monsoon in Nam, it was miserable at best
Soaked to the bone, impossible to rest
Extra socks in a baggie, try to keep the feet dry
Jungle rot rampant, till the day we die

First Lieutenants show up, they answered the call
Green and untested, but they know it all
Sergeants draw cards, to seal their fate
Who takes the lieutenant, on their first date

Replacements were needed, they weren't prepared
Bright and innocent, they thought their country cared
Welcome to the Delta, no one knows you're here
Back home, they are partying, smoking joints, drinking beer

A footnote in history, there is the real shame
These are folks we grew up with, politicians to blame
Skin diseases and rashes, hard to escape
Real World heroes, don't need to wear a cape

Got sent to the rear, every once in a while
A break in the action, but you won't see a smile
The Orange colored agent, often invisible for years
Taking a toll still today, leaving thousands in tears

Immersed in a world, every step unknown
Low softly and silent, Charlie's kill zone
Rice paddies laden with danger, look for the signs
Mosquitoes and leaches, trip wires and mines

The 'lucky' ones came home, to taunting and abuse
No solace to be found, many turned to the juice
They deserve better, for their huge sacrifice
Gave everything they had, in the land of the rice

Better late than never, let's start today
Let them know they're valued, find the words to say
No matter where they go, how far they roam
We honor them all, with a huge WELCOME HOME

- *Mickey Johnson*

VIETDAMN

It was the 60's, and Bedford was hot
Rock and Roll was big, Bedford Lake was the spot
Weekends would find, the teenagers all there
Guys flexing their muscles, girls brushing their hair

A battle across the globe, in another land
The Invasion was starting, band by band
The Beatles and the Stones, were part of the charge
Wouldn't be long, before they loomed quite large

Another battle brewed, Southeast Asia the place
The North against the South, The US set up a base
Most folks never heard, of a place called Vietnam
Didn't care what happened there, couldn't give a damn

Cambodia was hot, no business there
Just my opinion, we didn't have a prayer
We wasted lives, taking Hamburger Hill
Ordered to withdraw, those men died for nil

Roger Goff from Big Island, Bedford's first loss
Then Alfred Robinson, the next to cross
Michael Moorman from Thaxton, died there next
Then Tucker Oliver, Bedford feeling the effects

My teacher's husband from Moneta, Elmer Nance
George Davis from town, he didn't have a chance
Medic Lonnie Parker, one of Taylor Mountain's best
Eddie Daniels joined his brother, as they laid him to rest

Kenny Fink died back home, from wounds he received
The infection was too much, his family grieved
Montvale's Wayne Garrett, and Allen Tyler too
Stanley Taylor from Boonsboro, part of this sacred crew

John Samuel Myers, another Moneta young man
Bronze Star with Valor, for taking a stand
Maybe we could thank them, and remember the sorrow
For these young men from Bedford, there is no tomorrow

- Mickey Johnson

TRENDING TOWARD NORMALCY

For LLF - folks and friends and family

suggest I have PTSD
maybe possibly probably
I know I'm still debriefing
Nam and street cop coping
family scattered into a past
present and future I'm
retired into comings and goings
ok with certain disabilities
hearing loss and Agent Orange
told by many go to the shrink
take all the so-called remedies
no no no no
I tell each and all

kiss off

you my dear friends
(shrink if overhearing)
cannot understand my metaphors
my alter egos for conversations
my singing and kinships
with special beautiful strange people

I strive to accept perpetual debriefings
move on writing poetry to enrapture
every pre-post-stress syndrome
of all our hearts and minds

should I feel a need to scream
I'll just scream and move on or
laugh and cry and love
make time to pray or meditate
lay me down at night
seek to slumber in a dream
my enchanted forest

- Richard 'Eric' Johnson

THE WHEN AND WHERE OF NOW

in the interludes I make time
share some love sing some songs
paint pictures write poetry
go about routines or venture a new journey
while the continuum continues that then this now
like a commercial jet engine takeoff thrust of power
versus the fighter bomber takeoff booster
a traffic helicopter versus a Huey chop chop chop
4th of July aerial bombs and war zone cannonades
Mekong saturation bombings Frisco earthquakes
New England rains monsoon horizontal stinging water
heat humidity insect mosquito slithering snakes
comparisons crescendo on and on into all the senses
hearing disruptive noises imagining silence without fear
seeing horizons of promise or violence
smelling flowers and smoke and tasting cultural cuisines
touching touching touching anything anyone
observing always scenes of carnage or bliss and contemplate
a child's cry raising intense heartfelt concerns
and my dreams shake or soothe my heart awake
soul I can write poetry and sing to myself always knowing
a rediscovered memory is always an interlude
but never is the memory new

- Richard 'Eric' Johnson

BROKEN UNDER PRESSURE

Weight of the past
"Normal" for some to bear,
a feather of a weight,
unlike the pack I carry.

Am I alive inside?
I live like a ghost,
Unseen, denied peace,
beneath this veil I wear.

How do I return
back to solid ground,
to the shelter that holds
the truest parts of myself?

It eventually burns,
it all perishes.
It will transform
stronger, more resilient.

Will I mold into a treasure,
or will I crumble under pressure?

- Chris Kallas

RED AND BLUE

Well they served their country
But the country didn't serve them back
Their brothers and sisters turned away
Showed their back
Suckers and losers
They were red and blue

"Forgotten War" and forgotten warriors
Baby killers and worse
Hide yourself any way you can
Alcohol, drugs, anger and pain
Twenty-two a day is the price they pay
They were red and blue

But some stepped up
To help a friend or a neighbor
"Welcome home" and a helping hand
Because they served this land
They were red, white and blue

- Michael K. Lennon

COMBAT LULLABIES

In the killing rhythm of my dreams
Bullets pour like syrup from my M16
Or I spatter the hapless enemy in half
Or I'm trapped in a bunker, or chased
By VC I can't escape.

One night I said to a doctor
Himself unscathed, "What do you know?
I was the medic. They all came to me."

And one night, the platoon, ridding itself
Of old grenades, tossed them close to
Where I lay. Stop! I yelled, then
Hugged the earth as frags sailed past
Waiting, as we did that day
Wilson, last on the piled men
Hit worst by the Chicom's blast.

In the dream
Ours did not explode
But still, before the shrapnel storm
I knew would come, Stop! Stop! Stop!
I begged them, the words like rockets
Bursting from my youthful mouth.

- Marc Levy

QUAN LOI

Like many grunts
I laced my dog tags in my boots.
Soon enough
Dented, scratched, bent
Locked in the curl
Of my blood types circle,
Crimped in my non-religion–
Engraved upper case,
Sheltering in the sum
Of my service number
Even my name retains bits
Of crimson dust.
Day after day, patrols
Jungle, ambush, monsoon
Our blood and sorrows
Our countless steps
Become our mud-laced epitaphs.
Etched in Quan Loi's dusty red.

- Marc Levy

I REMAIN

I carried the weight of steel and sky,
boots pressing into earth as if to remind me—
I am not apart from it,
but of it.

In the mountain nights, stars pierced the black
and I felt their light thread into my chest.
Perhaps the universe borrowed my eyes
to see its vastness.

So when my body broke,
and danger whispered like wind through torn canvas,
it was not I who endured alone.
It was the mountain holding my spine,
the river tide repeating: *rise again, rise again.*

Coming home, the silence was heavier—
once familiar rooms felt foreign,
the rhythm of clocks unnerving.
But even then, the trees
leaned closer with their green patience,
reminding me:
a branch grows stronger where the storm once bent it.

We speak of perseverance
as if it is muscle or will,
yet I have found it in stillness,
in the simple act of breath,
in knowing the universe has always
been breathing me.

If I endure, it is because
I was never just one soldier—
I was the field, the sky, the turning stars.

Even in loss,
I remain

- *Steven L. London*

SPACE ODDITY (WHAT'S NORMAL)

I'm as real as a dough nut
Ground Control
To Major Tom
Children sit
On the stairs

Ground Control
To Major Tom
Take your protein Pills
Count down is
Put your helmet on
1
2
3
4
5
6, Dark black burning holes.
7
8
9

10

Lift off

Bombs away
Complete
Obliteration

Ground control
This is Major Tom
It's a beautiful show
Like seeing the sun behind your eyelids
Ground Control
This is Major Tom

Planet Earth is yours
and theirs and there's nothing
we can do
Wash it down
With some strawberry milk

- Nicholas Lopez

CORRESPONDENCE

(1968 -)

Loving you was easy going,
Knowing when to step or step back.
What to ask to set day's tack,
Bending paths to one wind's blowing.

Knowing me and then foreseeing
What would follow or be found,
Ups and downs, turns around,
Brought the diamond into being.

Decades later here we stand,
Dreams we had been unwilling
To assume their fulfilling,
Evident beyond the war land.

Love, we prayed, to chart the way,
Not the mind or clay or day.

- Michael Lund

CORRESPONDENTS

RVN, 1970 - 71

War words, sad words too numbing,
Added onto every page
Sent to those who have so wanted--
Story pool, news, homecoming.

At the early springtime burial
Public praise and private sorrow
For the friend who has fallen.
Name to go deep in granite wall.

Uniforms on those who stand
Tend to every moment's honor;
Bugle notes sound far beyond
Town and deep in states within.

Fifty years we say the name:
Steven Henry Warner

- Michael Lund

COMING HOME IS A MISSION TOO

The jets roar low—then skies go still.
The duffel's dropped, the air feels thin.
A welcome sign, a choked-up grin,
But something's missing none can fill.

You trained for war, for storms, for sand—
Not traffic jams and dinner plans.
To fight is one thing; to come home
Another mission, fought alone.

What uniform for this new day?
What rank for grief you can't display?
Yet bit by bit, the pieces land—
You build a life with calloused hands.

A father now, a neighbor too,
You don't forget, but you renew.
You teach the kids to tie their shoes—
Each knot a bond, a war subdued.

The battles change, but not your code:
To lead with honor down each road.

- Mike Lynch

STILL STANDING, STILL SERVING

A badge once pinned, a vow once sworn,
Still beats beneath a coat well-worn.
Though boots are shelved, the march goes on—
In hearts of those the war has drawn.

No longer on a base or line,
But in the job, the clinic sign,
The classroom chalked with lessons learned
From nights where distant fires burned.

Resilience hums in every task,
A courage no one needs to ask.
It's in the eyes that saw too much,
But hold their ground with steady touch.

They do not seek a hero's crown,
But lift the ones who feel cast down.
Still serving, though the war is past,
Still standing tall, built to outlast.

Their greatest fight? To live and give—
To show us all what strength can live.

- Mike Lynch

STEEL BENEATH THE SKIN

They never see the quiet war—
The one that waits when duty's done.
Steel helmets off, still bracing for
A fight not waged with blade or gun.

The desert dust, the jungle's scream,
Now echo in suburban light,
Where morning rushes lack the seam
Of meaning forged in fire and fight.

Yet still they rise—each working day—
To teach their hands a softer trade.
To greet the world in suits of gray,
While scars in silence lie unpaid.

You'd never know the weight they bear,
How sleepless nights still draw them back.
They walk through crowds, unseen, aware
Of peace won on a war-torn track.

Resilience is a quiet grace:
A nod, a job, a child's face.
Not medals gleam, but acts that prove—
The soul of one who still must move.

- Mike Lynch

CORPS TO THE CORE

A wild crazy life
Needs to be tamed
Sign the papers
Seek the change

Sleepless arrival
Hustled, harried
Reclamation project
To this I am married

Mental and physical
The challenges mold
I'm becoming something
This I am told

Here is the day
We have arrived
Families watching
The marching with pride

New roads are traveled
Continue to learn
All life's lessons
There's rank to be earned

Compassion is found
Deeper than home
You learn how to lead
You continue to grow

All this instilled
Will serve you so well
You never forget it
And live life to tell

Eagle, Globe and Anchor
Always displayed
It runs through the blood
It courses the veins

It beats in the heart
The Corps to the core
The life it touched
Goes on to touch more

Forty years later
You look back and see
It all came together
Through God and the Marines

- Joseph Maslanka

MY CAPTAIN IN THE MIRROR

Celebrate as a Phoenix rising like a sage,
not your broken-winged pigeon on display.
Stop leaning upon worn crutches backstage.

Your voice softly weeps from its closed cage,
a mouse squeaks while the still wheel stays.
Celebrate as a Phoenix rising like a sage.

A butterfly may become a bird on the page
the way hands create an eagle out of clay.
Stop leaning upon worn crutches backstage.

There's numbness in your stare. I'll wage
your harshest judge still speaks every day.
Celebrate as a Phoenix rising like a sage.

Imagine seas with a lazy moon—the gauge
numb with neglect. Inertia leads to decay.
Stop leaning upon worn crutches backstage.

Stand tall—the time has come to reengage
and shed the sullenness of your dismay.
Celebrate as a Phoenix rising like a sage.
Stop leaning upon worn crutches backstage.

- Richard L. Matta

JUNGLE BOOTS

Jungle boots tread worn thin.
Part of me, my second skin.

They carried me to parts unknown
and never failed to bring me home.

Caked with mud and canvas torn.
They hate polish; they must be worn.

In deserts dry or jungles wet
when huddled 'neath mosquito nets.

Jungle boots protect my feet
on humid days in searing heat.

In medic's tent all alone.
Surgeons sawed through flesh and bone.

With laces frayed and eyelets gone.
My jungle boot; returned me home.

- Harry Mayer

LITTLE JACK

Little Jack Duffy went off to war.
In 68 he went back for more.
His stature was slight and his courage immense.
But his bravery and valor were just a pretense.

His pain was unknown, repressed and unseen.
Seeking solace in bottles and drugs was his thing.
All the psychiatrists, doctors and men
couldn't make little Jack better again.

Life in the Navy was all that he knew,
and when it came to an end, he was dour and blue.
Retired, alone, with a bottle of rum.
Last night Little Jack swallowed his gun.

- Harry Mayer

THE COST

War softens the fabric of society. To win, you must push them past that fabric to the edge of humanity. "Win at all costs" is a cost few are willing to pay.

You must destroy that fabric by taking out infrastructure and many of their lives. Soldiers receive the call to do what they do while wishing they could go home.

It's a cost of life in the name of death. Where the decision of life is preserved by death. Death for a cause that so many support, but in life too many don't fall on their knees.

In this congregation of life and death, it's either you or them. Shoot, move, and communicate; it's the only choice you have. If you walk away, they shoot you in the back.

In this bad picture show, where few will ever go, in combat, if you are lucky to survive, it leaves you with hollow eyes. In this club only those who've experienced it understand looking at life through war-torn hollow eyes.

- Richard L. McClellan

A LEAP OF FAITH

boredom – the danger
of many young folk, some seeking changes
they took to singing

but their voices made
dogs howl and friends scowl, leaving a
hard choice lingering

Army Rangers! thrills
beckon, give us a parachute and
we'll leap into the fray

then training starts and oh
this sucks, as thrills morph to sweat and pain
a toll they must pay

I'm hungry! they cry
as veterans retort – silly soldiers
you ate yesterday

months later few remain
now Rangers, they rehearse every day to
breech the enemy's lair

still a leap of faith mocks
reason, jumping from perfectly good planes,
into the night's cool air

packs hung under the reserve,
chute, forty pounds of explosives inches from
naughty bits and underwear

so when you next greet a
Ranger, clap, cheer, just don't ask about their kids
unless they volunteer

and as these new Rangers
drink and sing, dogs still howl, friends scowl, but
no one dares interfere

- Sam McClintock

WOOD AND STEEL

When the Navy washed its hands
of me I'd been working in Supply
on shore but I found no bridge
to safely return to civilian life
The maturity service supposedly
tattooed faded and got deep-sixed
I drank true to the sailor cliché
and reverted to the deckhand bottom
ladder rung where my hitch began
I painted houses and often
a couple of sheets to the wind
on a forty-foot ladder I bounced
off clapboards mind swinging
in a bosun's chair hit or miss
globs of haze grey on either
mast, main or fore, machine or deck
shades on other ship skin
A kaleidoscope of tones smeared
my pea coat that came
to resemble drippy abstract art
My sanding was half-assed
Curling tongues of steeled
cheap glops mocked me
The bristles on the under a buck
four-inch brushes loosened and left
trails of hieroglyphic graffiti
I swear I'd seen HELP
The boss had no truck
with Navy neatness
just obsessed with all ahead full
wrap it up and scram to make
another shamble short term new
When I dragged myself
to a recruiter who landed me
on a Fast Frigate I was billeted
in Supply and not Deck
It was as if word of my oak, pine
and spruce coverings had traveled far
and wide like an SOS from a craft lost
at sea but I logged smooth sailing from
then on, Paint-by-Numbers ease

- Tom McDade

FANCY WORK

I saw a Bosun impressing
a fresh out of Boot Camp
addition to the crew
with some tricky knot tying
One-handed he weaved bright
red shotline civilians might call twine
around three fingers of the other paw
The Turk's head was fancy work
decor for Quarterdecks and dolling
up captains' gigs and things ceremonial
I'd mastered all the knots needed to sew
on three stripes to rise from apprentice
to full fledged seaman yet I grabbed
the challenge fingering and failing
on lookout watches forward and aft
and in my bunk as if praying a rosary
My first success struck nearly unawares
while I ogled the Argentine Libertad
its wealth of sails unfurled on the way
to a landmark Tall Ships event in NYC
From starboard amidships
I tried to pick out any armament
it carried and I might have imagined
swashbuckling movie cannons ablaze
if warships had not begun bombing
Vietnam a month past - all I knew
of war was one-sided and harmless, passing
powder kegs and projectiles to nourish
a 5" 54 during faux sea battles
In that deafening magazine I'd done
dud Turk trial runs so many times
a victory seemed thousands of nautical
miles away likely matching the distance
Virginia Capes to Vietnam?

- Tom McDade

ICICLE RIVER

Early into the gorge, I edge down
to the snowmelt fed river clearer
than air in calm spots and soda bottle green
where ripped by the stones it pushes over and around.
Boots off, I tenderfoot to find my place
on a copper, silver, gray green boulder inside
the current. A carpenter ant approaches
and with some persistence is persuaded off its path.
A butterfly dips and bounces on the current of the breeze
while the wind breaks over and around me in currents
mirroring the river's course. The trees laughing
swaying, nodding, welcoming; I am humbled
by the grandeur as time extends
its gift of now
          and now
                    and now

Pine siskins kiss the morning and I catch
a glimpse of God's shadow fleeting at the tree line
and on the reflecting water.

In this, I am comforted to know
these rocks will outlive us all.

- Julie A. Niven

THE UNIFORM

We start early in our efforts to be more.
We steal Father's undershirt to craft
into a superhero outfit. We sneak Mother's lipstick
to garner the power its donning appears to bring.
We seek role models to emulate, strive toward,
strike a pose in imitation of. We are taught
about the more of God if Christian or Jew
and the more of many others if other than.

And then the uniform – there's more there
than meets the eye - though that in itself
is a large part of it. Inside is where the real
synergism happens. I put it on and am more - in my mind
and in the colluding minds of the collective most
the uniform is a symbol of more – making up
in some respects for a neglectful or abusive childhood,
poor school performance, the hunger for more and the belief in
something that might possibly save us, protect and give purpose
to a world miserable in its search for meaning.

The uniform effectively serves
to quiet the questions and unspoken
sense of purposeless, guilt, shame, complacency -
close as the air we breathe
just under the skin.

The uniform undoes the gray
dichotomizing the world into black
and white, right
and wrong.

Wearing it, I become we and we
stand for what is right
and I am
and can be
more than I was before
in the uniform.

- *Julie A. Niven*

LIFE'S SKINLESS PONCHO

Another pullover of pain,
I get dressed to unveil the scars.
Why couldn't we find the moment to be us,
Away from the winds of separation, things would have been different.

The hurt stays repressed,
It finds a way to the surface when needed most
Like a worm squirming out after a rainy day.

Who cares about the lost souls that try to block my path,
They are obliterated by the fear of being me.
The boulders of loneliness crush them to pieces forever,
Roadblocks of flesh change direction.

- Jason Norton

PUSHING THROUGH

Strong arm the breadbasket, remove the window coverings,
Play games in the street while buses exhale their poisonous fog for free.

Playgrounds rumbling with tired rusted chainless swings,
Same hang out with dim flickering streetlights reminding you of tomorrow.
No utensils mock the caveman.

Leaves falling without warning or sound,
like nature's colorful Kamikazes.

Inkless pens replaced with a feather,
announcements made of sticky flypaper replacing your damp handkerchief.
What comes now without frequency?
Know now who you are,
Who do I recognize in these faded parking lines?

Let me find a new face or place I can help,
Distortion from wrinkles and unrest, how do you leave if there was never a way out.
No indication of return or warning.

Winds of vengeful peace submit their choices of broadcast,
Feel my hands in shallow glass puddles.
Mirroring parts of me to make a whole.

- Jason Norton

ROUNDTRIP

These words protrude out like scattered weeds,
From crevices not recognized known until now.
If pulled they will survive, and roots sprouted again,
No poison but love can destroy their place in the cracks.
Dirt and dust are their friends,
Brushed aside with the wind and passing footsteps.

Sunlit abrasions of guilt manufactured by disgust of what ifs.
Permission to revolt again and again,
Life's fence left many scars, it seems higher and higher to climb.
Round and round so many faces,
Who is here for the right reasons.

Gears need re-greasing,
Proper maintenance required.
But where to stop,
Where to start.
There must be a way off this road.

- Jason Norton

EMPIRE OF ECHOES

Twenty first century new wars
Same as the old except on different shores
It's how empires and upstarts keep score

We play back the two world wars
To us victory seemed assured
It was not even though we endured

Could've been the man in the high tower
An entirely different encounter
Germany and Japan the superpowers

Cold War turned on economics
Gorbachev's Perestroika and politics
Yeltsin had to weather the final Soviet apocalypse

A boon for KGB practitioners
Topflight diabolical grifters
They looted the country like shoplifters

Fast forward to the current United States
National debt going to seal its fate
Like an empty church collection plate

China waiting in the wings
Mao's revolution finally ready to sing
The world wonders what that might bring

American empire at a crossroads
Not farfetched it could implode
Following chaos would explode

Past deeds may come back to haunt
Long held hatred manifestly wrought
In the end it's our own damn fault

- Matt Overs

NUCLEAR SHADOWS

From Hiroshima to Nagasaki
What will the ultimate end game be
We'd just have to wait and see

Rosenbergs executed for treason
Welcome to the nuclear war season
The only thing we have left is reason

The Cold War begins unvarnished
The next forty years are tarnished
Military power being harnessed

A master arms race to doom
In a matter of hours, it could all go boom
Watch the video of that desert mushroom

The red scare witch hunt ensues
McCarthy conducts nationwide interviews
Hollywood studios subdued

The domino theory emerges
Leading to Korean and Vietnam excursions
Cuban missile crisis comes close before divergence

Fallout shelters become standard
School kids under desks once seemed outlandish
Movie The Day After meant to scare us

In the end the silos stayed closed
Negotiations led to reduced loads
No one had to use nuclear codes

Now the threat is very different
North Korea is largely still ignorant
It could lead to a predicament

Dark clouds seem ever constant
New battles can spark in an instant
Long term peace remains very distant

- Matt Overs

AIN'T THAT CLARENCE'S BOY?

Ain't that Clarence Parker's oldest boy?
the tall one, skinny from farm toil
sun-baked from priming 'baccer and soy
plowin' 15 acres of soil.

Marched right to the War Office, I hears
for Uncle Sam, he put his pen
'twas against his mama's say and tears
She knew some don't come home again.

A goodbye hug, then shipped overseas
to war's horrid, death-sifting sieve
to bring Nazi armies to their knees.
He prayed, "Don't let my Ma grieve,"

In France whizzed shrapnel, fear, and frostbites,
mud, then ice and the ragged sleep
in the crude foxholes between the fights
in the Alps, with the snow knee-deep.

With minefields, snipers, and mortar booms
that congealed into one endless Hell,
he just accepted fear and doom,
knowing sulfur was victory's smell.

It read, "Meritorious Action"
that official Army award,
describing only a fraction
of how they hugged them who'd warred.

Mama exhaled, "Oh hell with war!" –
She prayed, and the war was done:
She prayed for her boy and the Army Corps,
God brung most 'em back with her son.

She heard a car honk up the lane,
Looked up and tears fell on her dress
"Ma, I am back, I am home agayn!"
She hugged her boy, said, "I'll be blessed..."

He left a lanky farm boy, bone-limbed
come back home a strong straight-backed man
Musta' had somethin' deep inside him
Despite fears, he took a stand

Yep, that's Clarence's boy, I take it.
'Think I'll grab his hand and shake it.

- Jeffrey Parker

ONE SOLDIER; ONE FREEDOM

In honor of the US Colored Troops who joined the US Army as soon as they gained their freedom following the Emancipation Proclamation in 1863.

Who chopped that sprouting sapling of a tree
before it could sow this new nation's promised liberty?
Leaving some enslaved, while others lived free.

Jefferson's yellow'd parchment talked about a new, great nation—
Such eloquent Revolution, while Monticello moaned for its liberation.
"Freedom!" – while enslaved plowed another man's plantation.

Just powdered-wigged words, just ink on a yellowed page…
While one man, stripped of a surname and still trapped in another man's cage,
Hungered for freedom, holding generations of reasons for rage.

After Lincoln proclaimed in 1863, one man signed with the Union Army
What shallow freedom he would have, if it left behind those not yet free

What is the price to pay for one house, one nation?
One soldier chose to fight, readied to die, for others' liberation
A seed must die, bury its life; only later sprouts salvation

Disease and enemy spies stalked him, bitter cold stung deep into his bones
The devil whispered, "You got your freedom, and pay, so now just go home"

The soldier thought – of all who still ached for freedom, day and night
How somebody's got to brave the organized, legalized hate – to fight
So my children will be free – to think and dream, to read and write

To free my children from cruel boots, and whips, like devil's claws
Free them from history's repeated, sharp-fanged maws

To own their own land, and hold their heads high
Free children of God, free to laugh a lot more than cry
Free citizens who vote and voice, and freely ask "Why?"

He was only one soldier, dressed in courage, and Union blue
Fighting for freedom's claim – that is now passed on to you

- Jeffrey Parker

A LOAD OF WAR

Beneath the load I grumble, sore
Carrying a load imposed by war
I accepted it quite unaware
Of the time it would linger there.
So light at first, I could not know
How through the years, it's weight would grow.
When caring helpers offer aid
I turn away, too afraid,
Of removing the mask I always wear,
To hide the scars I daily bear.

Beneath the load I tremble, scared
My soul's frailty openly bared.
Stumbling forward breath by breath
Afraid of breaking, but not of death.
Keeping my eyes fixed on the skies,
To which I someday hope to rise.
Where, in faith, I'll turn in trust
To Him who sacrificed His Son for us.
And then I'll cast my torment down
And see it powerless on the ground.

Then, what the war has placed on me
Will be graciously lifted and I'll be free.
To soar unfettered through the air;
Calling others to join me there.
To feel the wings that God can give,
And, in His peace to truly live.
This load, for years has beaten me down,
But I glory in the hope I've found
And know that God will raise me high
To shout my victory, from the sky.

- Gene Pearson

FIREWATCH

breathing the dark skies
while hidden by the desert
sparkles flying by

a night-time routine
a mission in completion
in the name of peace

a late winter month
a chest full of foreign sand
winds that beat my soul

bloodied and buried
silently screaming away
no chance of rescue

stuck in memory
one that replays every night
always, in some way

I still stand this watch
complying to old duties
dying every time

then the morning comes
respite in daylight hours
filling me with hope

- Juan Manuel Pérez

FIREWORKS

Whataburger Field
a nice, cool Friday night game
Hooks beat the Missions

stadium goes dark
the celebration begins
surveying spring skies

becoming transfixed
staring at each explosion
replaying my life

it was a cool night
back during that desert war
thirty-four years gone

planes were dropping bombs
watched it all from firewatch
lighting up the sky

a short time later
we drove past the aftermath
burnt bodies and trucks

a loud explosion…
I'm back in the stadium
my wife notices

- Juan Manuel Pérez

GULF OF TONKIN JOURNAL

Three years I spent upon its decks
We plowed the seas from dawn to dusk.
A thousand feet from bow to stern
a ship of war-a power in trust.

Within its skin five thousand live
and struggle to survive.
I think of home and all I love
I hope to stay alive.

We send our planes far over land
to targets mere points on a map.
They drop their bombs and return again
to reload and narrow the gap.

This war's not right this I can see
I say I'm just doing my job.
To help me forget I drink all I can
and try not to utter a sob.

We're sent off to fight this terrible war
which no one can truly explain.
We see all the death and wonder inside
are we all just going insane?

Day after day, night after night
when will it ever be done.
This grueling routine has taken its toll
can ever this war be won?

I'm hollow inside, my energy spent
I've journeyed to the depths of hell.
Someday I'll put this far far away
and wonder: will I ever be well?

The years go by, we're headed home.
Soon I'll be back in the States.
But I'm not sure I'll ever forget
the ghosts I know now await.

- Patrick Pilcher

A MILITARY CEMETERY

Honor is for the living,
you who stand upon our roof.
With flag in hand and marching band
there is no surer proof.

Beneath your feet, within these graves,
rest those who once cheered too;
with marching band, and flags in hand,
and gave their all for you.

'Tis no vocation,
safe
and kind.
(Few there are who are soldier-born.)
Yet determined band with flag in hand
have history, adorned.

Soldiers die in war / soldiers die in bed:
Ah, death is common enough !
But with living hand, we took a stand:
Epic legends were forged of such stuff !

Now, no rhythm is felt, we hear no sound;
we weep and laugh no more.
Your flag in hand and marching band
reach not this furthest shore.

Yet, we still may speak, beyond the grave,
and not in cryptic strains !
Oh you with waving flag so grand,
heed our eloquent remains !

Successor....!
These
Proud
Carved
Stones
(and flag, enduring !)
are your future's preamble, and guarantee !
Decorations wither and fade at forgetful whim,
but we are the testament of the cost to live free !

- Steven Pody

LOADED WEAPONS

Loaded weapons, we addicts are.
With killing ranges both near and far.
Equipped with bullets from the first beat of our hearts
Yet, in the beginning we were missing a few key parts.
A sight was added first, and then a grip so we could feel
But the next components were a timely chore
The Prince had to hone those from steel
A trigger to pull, a hammer to fall and finally a firing pin
It mattered not, where or how,
The suspense was all in the "When"?
So patiently he polished and oiled us, until the pressure built up inside
And then BANG!
We went off, and the question remains
Was it Murder? Or Suicide?

- Paul Anthony Powell

IN THE DESERT, IN THE DIRT

I am there, alive in Afghanistan
I walk there
With my brothers
Footprints in the desert, in the dirt

I am there, still in Afghanistan
My body lies there
With my brothers
Bombs in the desert, in the dirt

I am here, home from Afghanistan
Part of me there / most of me here
With my wife and my boy
Home from the desert, from the dirt

- Derek Praeger

DISFIGURED HELPLESSNESS

It was probably the brightest day in the valley.
Me: beckoned from my cave-dwelling
built into the side of a burial ground.
Free-roaming goats, liberated from
their killed masters,
weave hurriedly past
bits of exposed cadavers protruding
from the eroding embankments.

Abdul knocked.

The sun seemed blinding;
Villagers needed medical assistance.
They waited by the road
separated from perceived salvation
by twists of razor wire.
Two young boys, six or seven,
one, holding a listless infant—
jaundiced skin, festering wounds,
wandering eyes, opaque irises,
indifferent to the gleaming of the sun.

Through Abdul, they asked
for the doctor to save their baby brother—
Both parents slaughtered by Daesh
merely weeks before.

But the *War* takes precedence.
Infant sorrows too far gone.
Sometimes the maddening part:
knowing you remain as helpless as them.

In acceptance, they
slowly retreated to the *valley* below,
leaving the last memory,
of a flaccid, disfigured leg
bouncing in young arms.

The sun still shone bright,
as they renavigated the boulders
back to the indifference and
helplessness wrought
by this savage and distant place.

- Daniel W. Ross

EVOCATION OF ALL WE HAVE

The blood-drunk raven
recalls the Silver tower
of love and longing.
He, with strong-hearted passion,
wisely protects us from pain.

War raged brutally
beyond a gold wheaten steppe,
tucked under blue skies.
Savagely, we revealed love
amidst the frost of conflict.

I loved her before
sunflower seeds burst Russian
pockets underground.
And I will still cherish her
when everything blooms anew.

She was my shadow
raging against the sun's
blind oblivion.
She eclipsed me like the moon
no longer loving the light.

She filled an ocean
of what may or may not be
with hell-forged longing.
I sailed through rough seas yearning
for deep ports of safe haven.

Bedridden, I sift
through life's sweet melody;
every trill, each beat.
I smile, treasuring the vast
orchestra shielding my soul.

- Daniel W. Ross

CHOW IN HEAVY WEATHER

Waiting in the chow line,
Knife and fork in my shirt pocket
Feet splayed apart for balance,
Tray grasped firmly in my hands
Deciding what would set best in my stomach.
Meat? Potatoes? Green beans?
Maybe crackers, or perhaps an apple.
No bowl of soup this meal. No mug of coffee either.
No extra sloshing needed.

Cautiously I make my way to an open spot at one of the tables
And just as cautiously put first one leg then the other over the bench
Precariously balancing my tray
Finally sitting, with legs tucked under the bench.
Setting my tray down at the first opportunity
Pulling out my knife and fork, one in each hand
Quickly I pick up my tray as Westwind takes a hard roll
First to port, then to starboard.
Back to port, with a lurch to starboard.
I wait for that in-between wave to level things out.
Quickly set down my tray. Quickly shovel food into my mouth.

Then the next hard roll to port necessitates
Hoisting my tray. One. More. Time.
Chew and swallow. Wait for the next in-between wave.
Repeat the process. Grab my tray; hoist it in the air.
Chow in heavy weather can be a real pain.

- Dave Schlitz

REMEMBERING

We have grown old now
The body aches. The bones creak
The meds no longer work as they once did
Our span of days grows shorter
Already its next week, next month, next year
But sometimes unbidden, memories come
Sitting in a rocker on the front porch
Dozing as early evening shadows
Slowly seep across the lawn
Snoozing in an easy chair
As the old dog waits patiently
For us to come and play with her again
A vague shadow crosses the mind
A fleeting image that slowly coalesces
Of our old ships, our old stations, especially our old buddies
As well as those we would have liked to have smacked
Up alongside the head with a 2 x 4…… Remember???

Guys we busted our ass with, got drunk on liberty with
Who always had our back as we had theirs
Especially those guys who went out when duty called
But never came back

Once more
Pushing a 40 Boat through heavy weather because folks need rescuing
Working on a Black Hulled Buoy Tender
Setting markers on highways of water
Sailing in a small ship on a vast ocean keeping a lookout
Banging through polar pack on a worn-down old Icebreaker
Knowing we would be back there in a heartbeat if we could
Wearing our dungarees and chambray shirts
With white hats cocked every which way
Always a little rough and tumble
Drinking, smoking, chasing after women

Looking back, it didn't matter if you were
Driving 40 Boats across the harbor,
Scraping buoys the Black Hulls.
Standing watch on the White Ones.
Sweatin' your butt off in a Breaker engine room
When the chips were down and the job demanded our very best
We were willing, and able, and Always Ready......Remember?

- Dave Schlitz

THE FIRST DAY

Evening, late May, 1961

A Coast Guard cutter floats quietly beside the Navy Yard pier
Down past the Hammerhead crane. Toward the end of the dock
I stand there in the twilight, gathering it in as darkness slowly descends
The tang of the waterfront permeates my senses, awakens my wanderlust

I am nineteen years old and think of myself as full of piss and vinegar
I am to report aboard the Polar Ice Beaker *WESTWIND*
Bound for the High Arctic where it will take me to
Places I have never been. Places where few have dared venture
Far away places with strange sounding names
Far away places seen only in movies or read about in adventure books
Soon I will be sailing to the very ends of the earth

For now, *Westwind* is tethered to the dock
Floating quietly, bathed in early moonlight
Snowy white from stem to stern, except only her mast and stack
On the bow, the barrel of her 5"38 gun pokes out
The port side bridge wing looms above me
A Monomoy surf boat sits nestled in its cradle
Just forward of two LCVP's tucked snugly one atop the other
Is the quarter deck, awaiting my arrival
Toward the stern, the flight deck awaiting the soon to arrive helos

My hectic day is almost over. Reporting aboard is all that's left
I hoist my seabag up and onto my left shoulder
Grab ahold of my AWOL bag with my right hand
Stride smartly toward the brow, stepping forward with my left foot
I go straight up toward the gangway, ready to step onto the quarterdeck,
I set the bags down, facing aft, I salute the National Ensign
Then the Quarterdeck Petty Officer
"Request permission to come aboard"
The duty Quartermaster of the Watch returns my salute
"Permission granted."
"Schlitz, Edward D. SNAG reporting aboard"
"Welcome to the *Wind*."
An entry is made in the Ship's Log Book
I am now part of the *Westwind* crew.
My grand adventure begins.

- Dave Schlitz

COVERED

As a result of leading Soldiers through four combat tours,
witnessing the darkest side of humanity,

and dealing with dashes past death's doorsteps daily,
I realized that I was covered in grief.

Being covered in grief has caused me to make peace
with my inner demons in order to find relief.

What is not said, daily I walk with the dead.
They will not leave me alone.

Sometimes grief even speaks.
It tries to cover and smother me.
It has even tried to chase me into traffic,
over bridges,
towards abuse,
and into pits of despair.
My grief wants me to think it is everywhere.

I am not trying to alter your beliefs,
but I have never seen good grief.

As a result of handling way too many body bags,
notifying numerous families of the death of loved-ones,
and presenting several flags to spouses,
I now understand why PTS stalks me.

Walks with me, talks to me, and tells me it has no equal.
I really do see dead people.

People who would be dead without the Army
and people who are dead because of the Army.

I smile through layers, hide my despair, cry through prayers,
because this part of the Military has no fan-fare.

I now know you cannot repay evil with evil because that just makes recovery too steep. You will never obtain peace.

I stand here as living proof that you can overcome
that gloomy covering of grief.

- James Sears

FROM THE HEART OF A COMBAT LEADER

Have you ever stood strong when fear was all around?
I mean, did you actually stand tall while others fell to the ground?
Have you ever done what common sense said not to do?
Even when others ran and said "come on, come on man, do not be a fool."
Has fear ever gripped you so tightly that you had to concentrate just to breathe?
I am talking about the kind of fear that brings even the strongest men down to their knees.
Have you ever had to remain calm and lead others to a safe place,
while fear was clouding your judgment and causing your heart to race?
Have you ever been so engulfed by fear that 48 hours passed without you missing a single beat? Heck, I almost went three days on just two hours of sleep.
Have you ever thought this could actually be the last breath I take? Because there was death all around you with no room for error, absolutely no room for a leadership mistake.
Have you ever been so frozen by fear that you just did not know what to do?
I have, but I had to go on because 237 Soldiers were looking at me to lead them through.
Fear in the face of leaders could destroy the will of many people,
while fear that goes unfaced could become deadly and extremely lethal.
Have you ever had to give an order that even you were afraid to do yourself?
I mean a command so crazy, you would not do it for all the world's wealth.
Have you ever had to send someone on a mission that you knew they might not complete?
That would be one of those nights when this leader did not get any sleep.
Have you ever had to give CPR to someone who worked for you?
I worked on Specialist Seirra for over one hour, but he did not pull through.
Have you ever had to carry the body of someone who saved your life?
The whole time I was thinking, what in the world am I going to tell his wife?
Have you ever had to tell someone that her loved one was no longer alive?
Trust me, you can actually see her heart breaking through the tears flowing from her eyes.
Have you ever stood over a person who was dead due to an order you gave?
Do not stand too close at the funeral because you may want to jump into his grave.
Have you ever had to stand before a grieving family? It is completely sad!
Their Soldier gave his life for his country and all I had to give them back was this US flag!

I am a leader of Soldiers because that is what I was called to do.
Just for the record, I am not trying to prove anything to any of you.
Most people just do not understand a true combat Soldier, warrior, and leader.
Here is some insight for all you poetry listeners and readers.
How did you feel when the Trade Center fell, the Pentagon got hit and those planes hijacked? This leader was devastated, thousands dead on my watch and it was my job to prevent those
attacks. Most really do not understand the sacrifices we have made for the armed forces.
I have missed birthdays, holidays, anniversaries and been through a couple of divorces.
We learn loyalty, honor, and respect for the US and for freedom. But we actually fight
for our fellow Soldiers, you would understand if you ever took the time to meet them.
We have faced fears for you and even for those that act like complete jerks,
because failure is simply not an option when Soldiers go to work!
I have faced my fears through multiple wars in the Middle East. I salute and thank those
Soldiers dealing with fear right now, so we can live here in peace!

- James Sears

NIGHTMARE IN PHU BAI

Out of the night, the fearful night,
In wave after wave so deep
They massed to the left, they flowed from the right
And slew us in our sleep.
The camp was lit by phosphorus flares
To show where the enemy lay
But their light, so high, just hanging there,
Revealed us as bright as day.
I thought of my home, my children and wife,
Of how I had left them too soon
To come to this place, a faraway life
As strange as the back of the moon.
I carried no weapon, no knife or sword,
My faith was my only reprieve;
A cross on my helmet, and God's Holy Word
Were all I had to relieve.
My fear was in charge, my courage was dead,
My faith to trembling gave way
As we crouched in our hole and covered our heads
And prayed for the light of day -
To live for a day, to live for a year,
To go home to my family and friends.
Each night was forever, a lifetime of fear,
A night that seemed never to end.

Out of the night, the fearful night,
With nightmares, sweat dreams and pain,
I wake in the light, afraid of the sight,
That I am back in Phu Bai again.

- Gordon Smith

THE D. I.'S HOMAGE TO EMMA LAZARUS

Give me recruits from broken homes.
Give me recruits who are abandoned.
Give me recruits with no place to go.

Send me these, the tempest-tossed,
the destitute, and the castaway, and
I will give them the family they seek.

I will teach them discipline
and have them marching with
the cadence of a true believer.

I will instill in them
the enduring values of
honor, courage, and commitment.

And after they are infused with
an unwavering allegiance to
the Constitution of the United States,

I will pin on their chest the iconic
globe and anchor of the Marine Corps
and tattoo on their souls a Devil Dog.

- Bruce Sydow

TACTICAL DIAPERS

Got pulled from a sermon on being humble
To change a diaper. Few things more fitting.
Kind of a life-sized illustration, that,
In the difference between humiliation and humility.
Humiliation you can get twisted, turned around.
There's those that like humiliating others and those
Those that kinda like being humiliated,
But nobody likes a dirty diaper.
That's service, there, and community,
Changing a diaper. Ain't never something you're doing for yourself.

Not that there's not pride in it, if you do it well.
There's technique and planning and preparation,
Having your kit packed out just right and combat drills.
There's few things that can go so wrong so fast
As changing a diaper the wrong way or at the wrong time,
And God help you if it's the wrong place.
Though I wouldn't mind not being the one to do it,
Ain't no point at all in waiting for somebody else to pick it up,
Even if your shift's just about complete.

It's a lot like war in that. And combat.
It's not that I necessarily want to be the one that gets the call
And wouldn't mind having a few less changes under the belt
But when it's time to get it all cleaned up
I'm fast and thorough and I'm glad to be of service.
But most important, most precious gift of all,
Is if I'm the one that's changing
And not the one as is getting changed.
Ain't nothing to look forward to in that.
And if you find someone taking pleasure in anticipation
You know they ain't never been there. Yet.

They'll learn. Lord, I'll teach 'em.
Ain't no shame in being a new parent,
We all had that glow once.
But if having changed a diaper or two
You decide that's something to aspire to,
Look forward to, bring home & wear around the house,
If you having changed a dirty diaper get so proud
You bring that diaper home and give it pride of place
On your kitchen table, well, I guess, then,
Guess I just ain't comfortable with you babysitting.

- N. Jed Todd

V, BUT NOT FOR VICTORY

The V in a medal, that doesn't stand for Victory
But Valor. And Valor is a euphemism.
What it really means is "Bad Day".
Means someone somewhere had the worst day of their lives
Out there. May well be the last one,
For them or some of their buddies.
Maybe just the ones they call the enemy.
Must have been someone's last, at any rate.
The medal isn't given as an honor,
A celebration, a call to glory or adventure.
It's an apology, a way to say we're sorry.
Sorry that we didn't plan it better,
Sorry that we took so long to reinforce,
Sorry that the MEDEVAC was delayed.
Sorry we weren't there to help
And keep you from getting that award,
That little bit of colored ribbon
For which Napoleon, misled, thought
For which we fought and died.
That ribbon isn't anything,
But the memories are all.
For those that never got a chance to wear it,
For those without a chest to pin it to,
And to whose sons and widows
We present
A V device on a bit of
Colored ribbon
But that V ain't there for Victory.

- N. Jed Todd

WE'VE GOT THE WRONG WORDS

Everybody always says to keep in mind
"The boys over there fighting for what we have here."
Hard to say what's more wrong. Boys, sure, that's old,
War does not discriminate. More than that, it's the places
That's wrong. God knows that ain't my war.

'Here' and 'over there' is wrong,
War is always here.
Immediate, immersive, all around you.
Once you've been you never really leave,
But you dream of over there.

Over there where you came from,
Or maybe where you hope we're going.
But that misty dreamlike vision, over there,
Never all that far, but never quite in reach,
Goal, memory, or promise. It's why we fight.

Some fight to keep here from coming over there,
Some fight to bring here into over there at last,
But either way it's *over there* that is the intent
Over there is why we do it, what it's for, and how,
Here is what we're fighting through to get there.

- N. Jed Todd

A WOMAN'S PLACE

My mother immigrated from Poland to America and
intended to live in California, but
fell in love with New York's
grittiness.

She, a single mother of two,
filled my life with prose and poetry,
facts and fiction.
"Knowledge is power,"
she recited to me.

I grew up being told,
no—expected—
I can do anything that my brother can do.

The idea of a girl's capabilities being different
or less than,
from a boy's is foreign to me.
Nonsensical.
Laughable.

As an Air Force Captain, I found myself teaching
air traffic control fundamentals to a group of young officers,
among them three Saudi men,
who believed my place was at home.
They told my commander they preferred
a male instructor.

Little did they know
my mother made me
fearless.

I said,
"It sounds like a them problem."
And I took my rightful stance
in front of the classroom.

- Catherine Torres

WHAT MATTERS MOST

I stand on the shoulders of women who came before me.
Those who asked to wear the uniform the minute they lawfully could.
Who petitioned to fly the jet,
Who wanted to see the front lines and fight
for the freedoms and equality that
supposedly is protected.

I can still hear the audible gasps in the audience, when I stood on stage in a
white, satin dress at a teen beauty pageant and announced
I will be attending the US Air Force Academy after graduation.
(I didn't win the pageant).

I stand on the shoulders of women who endured name calling by
small-minded men, who are accused of cheating for their successes.
The women who said, "They told me I couldn't,
so I wanted to prove them wrong."

The women that came before me
endured the brunt of it, so that my generation
could soar, less encumbered—
not unencumbered—because
small-minded men still make comments,
"She succeeded,
despite being female."
Despite?
They try to erase female accomplishments from history,
mask it as equality.

It's unfathomable that a female weapons system officer clocked
one thousand hours in an F-15E—a feat for any gender.
It's a wonder a woman's emotions don't hinder her ability to lead
a combat fighter wing, or a branch of the military, or anyplace that isn't home.
Will the Thunderbird team maintain its high standards with a female pilot?
What considerations does a woman need for a space mission?
What about her children?
(No one ever asks about a man's children).

I stand on the shoulders of great women,
and yet, the stars remain just out of my grasp.
Perhaps my daughter won't have to stretch as far.

- Catherine Torres

YOU AND THE TIME

I don't know what happened exactly, but here I am.
Motionless and forgotten
like a He-Man sticker
placed slightly crooked on the side of an old lunch box.
Throw me in the closet with the coats and the hats,
on the floor with the pumps and the flats,
or maybe in a cave with the ghosts and the bats.
What am I to you when all you do is move?
Does it look like I am in motion to you?
I hate to watch you fly by.
You know,
you and the time.

I caught a glimpse once,
but that was all I was able to catch,
besides a case of the heavies.
You know,
when you just feel…
heavy.

I've traded my idols for a life idle.
I've become intangible and oddly abstract.

But I'll move again,
one day though.
I'll be the most marvelous mover around.
I'll make the movers and shakers pause to bear witness,
to this peculiar and glorious occasion.
Yes, an unprecedented orchestration it will be.
I'll be better,
I'll be cured,
just like you
I'll be free.

As soon as I get these shoes all tied,
Assuming my feet and hands will abide,
As soon as I stand strong, sturdy and straight,
And push off this oppressive and soul crushing weight.
The weight of my fears all the years have assembled
while watching you fly by.
You know,
you and the time.

-Phillip Trezza

NORMAL

When it comes down to it
I do not know how to be normal
for the next thirty or forty years
how does anyone who's ever
worn combat boots for a living
take them off forever

I work a regular job at a desk in an office
My coworkers have no idea how intimate I am with war
They don't know I hear the call of a machine gun
every time the door rattles just so,
leaving me longing for that familiar pattern
Play it again
Or last year when a semi blew its tire on the highway
and the vibration shook the office walls
I stood at the window's edge, alongside everyone else
watching the scene on the road below
I've never heard anything like it, they said
and I agreed like it was something new
As if the boom hadn't vibrated through my lungs,
echoed across my memory, paused my bloodstream,
and stalled out my heart
As if my first thought wasn't, *It's been far too long*

So I returned to my desk
and kept right on working
Kept right on breathing
My blood kept pumping
Like someone who is normal
It's been far too long

- Ali Watts

THIS IS OUR ANTHEM

We sang songs as we walked.
Not any marching cadences, but real songs.
Ones we remembered from the radio back home.
We had terrible voices and sometimes
botched the lyrics, but that didn't matter.
One foot in front of the other and follow the leader.
That's how we did it every time we went out.
Every step further and further away
from where we came and who we were.
The only thing that kept us together,
the reason we didn't lose ourselves entirely,
were the songs. Even the ones I didn't like,
even the ones I didn't know at first, I had
memorized by the end. The end of our walking.
When we were trapped, surrounded,
pinned down, and outmatched. When we knew
the only way home was through the bitter darkness.
We circled up and faced out and whispered
hopes to give us the will, to give us the courage,
to never give up until we could go on no more.
There would be no coming back from this, at least
not in the way we wanted to return. We promised
to find one another on the other side of this fate.
And then, all together, we sang songs.

- Ali Watts

AUTHORS' NOTES

David Martine Aldridge is a retired army soldier and civil servant who served our nation for forty-five years. He had three tours in Viet Nam with the Infantry and was awarded three Purple Hearts. David has been a student/member of the Mighty Pen Project at the Virginia War Memorial since 2016, and some of his writings appear online in Virginia Commonwealth University's *VCU Scholars Compass Archives.*

Zach Allen of Nixa, MO, served in the army on the Tactical PSYOP Team 1061 during operations in Sangin District of Helmand Province, Afghanistan. Zach is an alumnus of Missouri State University who enjoys the privilege of life in the Ozarks with his family.

Anne Barlieb is a poet, wounded warrior mentor, and army veteran with fourteen years of service as an aviation and psychological operations officer. She spends much of her time attending workshops through veteran services organizations and mentoring retiring servicemembers at Walter Reed. She is a volunteer with Armed Services Arts Partnership (ASAP), Community Building Artworks (CBAW), and Dog Tag. She lives in Rockville, MD.

Corey Barnes lives with his family on a small farm in Hillsboro, KY. He is a marine corps veteran with three combat deployments to Afghanistan. Corey finds fulfillment in sharing the poetry he has been writing for over twenty years, and he has self-published a small volume of poetry entitled *Legs of My Land.*

Jeffrey Beck is a retired army soldier and a graduate student in the MFA program at Southern New Hampshire University. He holds a BA in English from American Military University. He lives in southern Virginia with his wife on a five-acre homestead. He enjoys relaxing with his chickens, reading and writing fiction, fantasy, and poetry, and writing a memoir about his combat experience in Iraq.

David Reid Brown is an award-winning poet and a retired Navy Chaplain. He currently serves as a high school government, Bible, and art teacher at Grace Christian School in Mechanicsville, VA. David is the author of ten books. The most recent include *Psalms From The Sea, Volume IV: Operations,* and *Relatables*, his second book of poetry which includes the three poems published here.

Randy Brown is author of the award-winning poetry collection *Welcome to FOB Haiku: War Poems from Inside the Wire* (2015). A ten-year retired Iowa Army National Guard veteran after one overseas deployment, he embedded with his former unit to manage civilian media in Afghanistan, May-June 2011. He is a three-time poetry finalist in the Col. Darron L. Wright Memorial Writing Awards administered annually by the literary journal *Line of Advance*. Among other publications, he edited or co-edited the anthologies *Why We Write: Craft Essays on Writing War* (2019), *Things We Carry Still: Poems & Micro-Stories about Military Gear* (2023), and *Giant Robot Poems: On Mecha-Human Culture, Science & War* (2024).

Kenneth Conklin is an author, poet, former technology industry professional, and army veteran with service in Vietnam. Born in Glendale, CA, he has lived in Daleville, VA, for the past twenty-one years. In 2023, he served on the Centennial Planning Committee for the Poetry Society of Virginia. He is the author of *NORVEL: An American Hero* and many commentaries published in newspapers and periodicals. A collection of his poems can be found in *The Zen of Ken.*

Lisa Wence Connors retired from the army and lives in Colorado. She facilitates creative writing workshops for the Mesa County Public Library and the Grand Junction VA. She is the Editor of *High Desert Cadence: A Journal of Veteran Creative Writing*. Her work has been published in the journals *Proud to Be, Alexandria Quarterly, Gyroscope Review*, among others, as well as several anthologies. She has won two Missouri Humanities Poetry Awards and four Gold Medals for Poetry from the National Veterans Creative Arts Festival.

Sabrina Davis is an army warrant officer in the Judge Advocate General's Corps. She earned her bachelor's and master's degrees in criminal justice from Purdue University. Her work has been featured in the Mighty Pen Project archive and *Outside the Wire*, a veterans literary journal published by Rutgers University–Camden Writers House. Currently, she is completing a memoir about her search for her grandfather, a Vietnam veteran, and lives in Maryland. Her writing often bridges service, motherhood, and legacy.

Steven Davis is a retired army nurse who deployed into conflict zones three times, including providing nursing care to servicemembers and civilians in central Baghdad. In 2025, he successfully published *Keeping the Stethoscope, Hanging up the Uniform!: The Curse of Combat Disability Retirement. A Predictable Paradox: Patriotism, Policy, and the Cost of Service After the War* is his newest book. Steven is the recipient of an International Impact Book Award for his memoir writing. Through his work, he hopes to help veterans successfully return to society, so they can "fit back in." Currently, Steven serves as an ER Nurse Traveler to offset the worsening nursing shortage across the country.

Vivian Delts is an army veteran, advocate, and lifelong poet whose commitment to serving others has shaped both their work and their writing. Drawing from experiences in military service, community engagement, and continued support of veterans, her poetry explores themes of transition, belonging, resilience, and the promises made to those who serve. She uses the written word as both reflection and witness, giving voice to experiences too often overlooked and challenging the invisibility many veterans face after returning home. Through poetry, she seeks to preserve stories, confront exclusion, and affirm the humanity behind service.

Michael Dooney is a former Army Special Forces soldier who deployed to areas in Europe, Africa, the Middle East, Central Asia, and Afghanistan. After retiring from the military in 2004, he worked as an embedded contractor for several federal agencies. Recently, he authored two military science fiction novels, and he is new to poetry. Michael holds a master's degree in technology management and enjoys drawing and designing physical products. He lives in Stafford, VA, with his wife.

Richard Epstein served in the Army Signal Corps assigned to Northeast Thailand, and he was a civilian field engineer at communication sites throughout South Vietnam and JUSTMAG, Bangkok, Thailand. Richard attends several veterans writing groups, and he has been a featured reader at the Navy Memorial, The Vietnam Woman's Memorial, the Orange Bear in New York City, the Kensington Book Store, and the Philadelphia Ethical Society. He is the editor of two veterans anthologies, and his poetry has appeared in *O-Dark-Thirty, DEROS, Incoming, A Common Bond, Veteran's Writing Project, Beltway Poetry Quarterly*, and the *Schuylkill Valley Journal*. Richard also hosts a Veteran's Venue at the annual Kensington Day of the Book festival, Kensington, MD.

Herb Evert "I was on the bridge that day the chain snapped,"recalling a life-altering accident aboard the USCGC Mesquite. This blend of maritime intensity and reflective storytelling defines Herb's work. A retired university educator and administrator, Herb served six years in the coast guard, including two years as a quartermaster navigating the Great Lakes on an icebreaker. Today, he writes from his home in Cottage Grove, Wisconsin. Whether through his magazine features or his two collections of short stories and poetry, Herb draws upon his diverse background—from the bridge of a ship to the halls of academia—to explore the undaunted spirit of those he has encountered along the way.

Leo Farley is an army veteran. He served as a Communications Specialist with the 52nd Signal Battalion in Can Tho, Vietnam from June 1970-71. After an almost sixteen-year stint on Wall Street, he gave up his financial security blanket to devote more time to artistic endeavors in the theatre as both an actor and director. He is a founding member of the 29th Street Rep Theatre and currently Artistic Director of LSMFT Theatre Group, both in New York City.

Cynthia T. Ferguson, PhD, is a navy veteran and mustang who enlisted in 1982 and first served as an Aircraft Electrician and Navy Aircrewman in her military career. Later, she served as a Certified Nurse Midwife and Forensic Nurse until her retirement in 2014. Cynthia is a martial artist with a sixth-degree black belt in Japanese Shorin-Ken, and she studies Shotokan and other martial arts. While currently living in a small 1700's home on a bluff overlooking the James River, Cynthia enjoys feeding numerous hummingbirds during their summer breeding season. She writes poetry, short stories and novels in her free time.

Eric Forsbergh is a Vietnam veteran, having served in the navy for four years in the Tonkin Gulf and Da Nang, Vietnam, where he worked on the flight deck as an aviation electrician's mate. He practiced dentistry for thirty-eight years in Reston, VA. He won the Poetry Society of Virginia's Edgar Allen Poe Memorial. He has published two books of poetry, *Imagine Morning*, and *This Mortal Coil*, the second being about DNA and its manifestations. He has participated in dental mission trips to Guatemala and to Appalachia, as well as performed volunteer work as a COVID vaccinator in Loudoun County, VA. In 2022, he graduated with a master's certificate in biblical justice from the John Leland Institute.

Shelley Gill grew up in rural Virginia and enjoyed writing from a young age. After trying college without much success, she joined the military. It served as a gateway to a vibrant life of purpose and possibility, where she could serve her country, and do something meaningful. She joined the Air Force in 1995 and had a vibrant career spanning twenty-seven years of active and Air National Guard service. In 2022, she retired and returned to Virginia to raise her son and spend time closer to her family. She also began to discover writing as an outlet to process and understand changes, challenges, and loss. Shelley hopes to help other veterans experiencing disconnection and uncertainty with a message of perseverance, hope, and unity.

Bill Glose is an army combat veteran and former paratrooper whose transition to civilian life has been marked by a penchant for extraordinary experiences. From trekking 1,500 miles across Virginia to participating in a world-record-setting skinny dip, he transforms his adventures into magazine features. As an author, Glose has published six poetry collections and a book of fiction, with his work appearing in journals such as T*he Missouri Review, Rattle, Poet Lore, Narrative Magazine*, and *The Sun*. A former *Daily Press* Poet Laureate (2011) and a featured voice on NPR's *The Writer's Almanac* (2017), he also spent three years as the publisher and editor of the literary journal *Virginia Adversaria*. His work has earned him the 2023 Library of Virginia Award for Fiction, the F. Scott Fitzgerald Short Story Award, and the Dateline Award for Excellence in Journalism.

Richard Hamner is a retired Navy Civil Engineer Corps Commander with twenty-two years of dual enlisted and officer service. Duty assignments included Naval Security Group, public works, construction administration, Seabees, and infrastructure management at Marine Corps headquarters. He holds a bachelor of science degree in Civil Engineering from Virginia Tech and a master's degree in strategic studies from the US Army War College. Today, Richard is a budding poet. He resides in Dumfries, VA, with his wife Margie, where they are now grandparents.

Robert Haynes is a poet and army veteran who served sixteen years as an infantryman with multiple deployments to Iraq. Originally from Tucson, AZ, he currently resides in Woodbridge, VA, where he spends his time gardening and restoring classic cars. He earned his MFA in Creative Writing last summer, and these poems mark his debut publication.

Anita I. Homes served in the Women's Army Corps (WAC), where she wore the Pallas Athena insignia on her lapel. She is a published poet and writer as well as an award-winning public speaker, fiber artist, and origamist. Her ongoing *Ars Poetica* series explores the art of poetry and the life of a poet. Holmes enjoys making good coffee and hosting workshops for writers and artists.

Wayne David Hubbard is a poet, educator, and marine corps veteran. He enlisted from 2000-2005 with service in Iraq and Japan. Born near Newark, NJ, he lives in the Shenandoah Valley and is the current Treasurer of The Poetry Society of Virginia. He is the author of two poetry collections – *Mobius: Meditations on Home* (2020) and *Death Throes of the Broken Clockwork Universe* (2022). His poetry explores culture and science through the lens of everyday experience and appears in various anthologies.

Josiah Jack was once an army staff sergeant medic but always "a poet—who believes that whether you're writing on construction scraps, Humvee hoods, or perfumed parchment—poetry is the healing voice of life within us all!"

Mickey Johnson served in the army as an airborne parachute rigger from 1974 to 1977 during the Vietnam era. He has published one book, *Mickey's Unfiltered Musings*, and he has three more unpublished. Mickey likes to spend his time at Smith Mountain Lake golfing, paddle boarding, and playing pickleball. He started writing poetry a few years ago and believes that it has served as a real blessing for him.

Richard 'Eric' Johnson is an army veteran who served in Vietnam and West Berlin before embarking on a career in law enforcement with the Marion County Sheriff's Department. He holds a BA in Germanic Languages and an MS in Education from Indiana University. A Pushcart Prize nominee, Eric has published five volumes of poetry, including his 2026 collection, *Watching Angels Dance By Candlelight.* His poems have appeared widely in print and online journals, and his literary contributions are now permanently archived in the Vietnam War Veterans Poets Foundation collection at La Salle University's Connelly Library. A resident of Arlington, VA, Eric continues to capture the intersections of his diverse experiences through his writing.

Chris Kallas is a marine corps veteran with service in Afghanistan who lives in Baldwin City, KS. His writing reflects his experiences with mental health, resilience, and the realities veterans face after service. This is his first publication.

Michael K. Lennon is a marine corps veteran and the son of a marine. Mike served from 1978-1998, including northern Iraq, helping the Kurds in 1991. He lives in Rockbridge County, VA, and currently is the Board Chairman for Purple Heart Homes, a nonprofit helping veterans of all generations.

Marc Levy served with the Delta 1-7 First Cavalry Division in Vietnam and Cambodia as an infantry medic. Born in Newark, NJ, he resides in Salem, Massachusetts. His work has appeared in *Litro*, *CounterPunch*, *New Millennium Writings*, and *The Westchester Review*, and he has been nominated twice for a Pushcart Prize. *Medic in the Green Ti*me is his website.

Steven L. London is a writer, editor, educator, and army veteran and Purple Heart recipient with service in Afghanistan. Born in Mason, MI, he lives in Bulgaria with his wife, child, and dog, and currently leads the Writing Center at the American University in Bulgaria. His recent projects include *Unwritten* (finished manuscript) and *The Borrowed Century* (novel in progress).

Nicholas Lopez is a marine corps veteran. He is curently a program coordinator at Missouri Humanities. His poems and art have been published in *Veteran's Voices*, *Haiku Journal*, *Line of Advance*, Volume 11 of *Proud to Be: Writing by American Warriors,* in parentheses, and *Hellbender Magazine*. Nick's paintings have been displayed in several group exhibitions in Missouri and Kansas.

Michael Lund is an army veteran. He is a native of Rolla, MO, and professor emeritus of English at Longwood University. He has written several novels inspired by Route 66, America's Mother Road, and small-town life in the Tidewater area of Virginia and North Carolina. As an army correspondent in Fort Campbell, KY, and Republic of South Vietnam, Michael has published short stories about his military experiences. Currently, he directs a free writing program for the veteran community.

Mike Lynch is a marine corps veteran. He wrote "Still Standing, Still Serving" to recognize the unyielding spirit of service that continues beyond retirement from military service. The poem reflects on the quiet heroic acts of those who continue to lead, heal, teach, and inspire their communities through the courageous and compassionate actions they demonstrate while serving in uniform.

Joseph Maslanka is a marine veteran, playwright, and ten year member of the Mighty Pen Project whose work blends grit, humor, and lived experience. His play *Dress Code* premiered at Virginia Rep in 2023, and his ministry in the United Methodist Church shapes much of his storytelling voice. A former member of the Righteous Outlaws, he has supported veteran organizations through music while grounding his life in family as a husband, father, and grandfather.

Richard L. Matta, originally from New York's Hudson Valley, is an air force veteran living in San Diego, CA. His poetry has appeared in *MacQueen's Quinterly, Red Rock Review, ONE ART*, and many haiku journals. He was recently nominated for a Pushcart Prize, Touchstone Award, and Best of the Net.

Harry Mayer is a print artist, writer, and navy veteran. During his military service, he functioned as a Master Explosive Ordinance Disposal Technician, a role he followed with over two decades of experience coordinating federal disaster operations. Today, his creative practice focuses on linoleum block prints and mixed-media works that explore themes of resilience, memory, and conflict. His art has been showcased at prominent venues, including New York City's Salmagundi Club and the Ellarslie Open. As the founder of Art Veterans Studio, Mayer is dedicated to empowering fellow veterans through creative expression and humanitarian art. His writing has been featured in the *VCU Mighty Pen Project Anthology & Archive, the Military Writers Journal,* and *Warrior Writers.*

Richard L. McClellan is a native of Yellville, AR. As an army veteran, his life has been marked by extensive travel, from Japan and Korea to Portugal, Italy, and across the breadth of North America. His diverse interests range from genealogy and motorcycling to the intricate crafts of crochet and knitting. An accomplished student of technology, he holds a bachelor's degree in electronics from SMSU and an associate's degree in electromechanical technology from North Arkansas Community College. Though a competitive billiards player for many years, a 2007 injury shifted his focus, leading him to discover a profound passion for poetry in 2011. Today, he is an active member of the Poetry Society of Vermont, where he continues to hone his craft.

Sam McClintock is an environmental scientist and cybersecurity consultant living in Williamsburg, VA. He has worked with the EPA, industry and government contractors, and served with the 2nd Ranger Battalion and Defense Intelligence Agency. His alma mater is NCSU.

Tom McDade is a poet, writer, and navy veteran whose first tour of duty began in September 1963, and he served on the *USS Mullinix* (DD-944). Born in Providence, RI, he is retired and lives in Fredericksburg, VA. His work has appeared in *The Columbia Journal, GloMag, The Manila Review, miniMag Literary Anthology, The Story Sanctum* and other publications.

Julie A. Niven writes poetry and travel narratives, as well as crafts pottery at home in Virginia. She has been published in various poetry journals over the last 4 decades. She retired from the US Public Health Service Commissioned Corps (USPHS).

Jason Norton is a poet and recent Kean University student pursuing a MA in English writing studies. He is a 3rd Infantry Army Veteran soldier with overseas NATO deployment service in Kosovo, Albania (1999-2001). He was born and resides in Trenton, NJ. His recent publications include the *NJ Bards Poetry Review* anthology (2022-2025), Princeton, NJ Library People & Stories *Our Common Ground* anthology 2025), Veterans Poetry Project's Anthology *Perseverance & Resilience* (2026). He is also an active fine artist creating in various mediums that include abstract painting, sculpture, print making, and photography.

Matt Overs is a navy veteran who was born in Atlantic City, NJ. A longtime boxing writer turned novelist and poet, Matt's debut novel, *Downbeach*, was published by Moonshine Cove in 2024, and *Rhymes for the Times*, his first poetry collection, was released as a chapbook by Bottlecap Press in 2026. Matt served four years of active duty in the navy as a cryptologic technician in signals intelligence. He now lives with his family in New Jersey.

Jeffrey Parker is a poet and children's book author who served as Navy JAG and a career with the Department of Veterans Affairs. He is a member of the planning committee of the Roanoke Chowan Literary Festival in North Carolina and supports nonprofit literary organizations and African American history in northeastern NC.

Gene Pearson is a writer, playwright, and occasional poet who was awarded the Army Silver Star and the Bronze Star for actions in the Vietnam War. He is the father of two sons and four daughters. He enjoys reading, writing, and a peaceful life with his wife of forty-seven years in Chesterfield, VA.

Juan Manuel Pérez is a Marine-Regulation Navy Corpsman (8404) who deployed to Desert Shield/Desert Storm with the 2nd Marines in 1990-1991. This Texas Public High School History Teacher, Native American Gourd Dancer, and former Migrant Field Worker, is the 2019-2020 Poet Laureate of Corpus Christi where he lives. This past April, he was nationally recognized and received the 2026 Marine Corps Heritage Foundation's Robert A. Gannon Award for Poetry at the National Museum of the Marine Corps in Virginia for his poetic memoir, *Thirty Years Ago: Life And The First Gulf War.*

Patrick Pilcher is a navy veteran. He made three deployments to Vietnam, serving as a guided missile tech aboard the aircraft carrier *USS Constellation.* He retired as a park ranger for the National Park Service and currently lives in Seattle, WA, where he writes on veterans and military issues.

Steven Pody worked for the National Cemetery Administration for eleven of his forty-one years serving the United States Government, and his poem here provides insight into that service. His other work includes two books, *The Panoptikon* and *The Flutter-By Two-Step*. He has also been published in some thirty-two printed journals and on the Internet. An air force veteran (1975-79) and a two-tour Peace Corps veteran, Steven has lived in Alaska, the Middle East, and Africa. He currently serves as president of Riverside Writers in Fredericksburg, a chapter of the Virginia Writers Club.

Paul Anthony Powell lives in Salem, VA. He served six years in the army as an airborne/air assault infantryman. Addiction led to the loss of his wife and children and landed him in homelessness. Recovery for the last twenty-three years led to the restoration of all things, and he is now the community employment coordinator for homeless veterans at a VA Medical Center in Salem, VA. His first poem was written while he was in the homeless shelter recovery program.

Derek Praeger is a native New Englander, now happily living with his wonderful family in Southern California. He served honorably as an Intelligence Specialist in the Marine Corps Reserve, for eight years, achieving the rank of Staff Sergeant. During his time in service, he deployed overseas to Mozambique, Africa; Helmand Province, Afghanistan; and Pohang, South Korea.

Daniel W. Ross is an army veteran. He hails from Baltimore, MD, where he developed a love for poetry and creative writing at Gilman School and Goucher College. His poetry has been published by *As You Were: The Military Review*, *Line of Advance, Mouthful of Salt, Plants & Poetry,* and Washington *Writers' Publishing House*. He was featured as the artist of the month by *Artists from Maryland* for May 2026. He currently lives in Germany with his wife and daughters.

Dave Schlitz is a military veteran and author known for his memoirs and poetry detailing his time serving in the Arctic and high seas with the navy and coast guard. He is an active participant in the Mighty Pen Project, a highly regarded veteran's creative writing program based in Richmond, VA, and his memoirs are archived by Virginia Commonwealth University's Compass Project.

James Sears is a retired Army Officer from Louisville, KY, and he is a graduate of Louisville Male High School, Western Kentucky University, and Central Michigan University. He is a veteran with four combat tours who is a student of history, a lover of writing, and an advocate for veterans. He is best recognized as James "Mr. Speaker" Sears. Currently, he is a Department of the Army civilian where he supports the military while mentoring the next generation of leaders. James is the author of several books, including *The Heart of a Military Leader* and *Love Verses Hate*.

Gordon Smith is a retired public school teacher and former Navy Chaplain living in Jonesboro, AR, in assisted living. His wife, Carol, is also in Jonesboro, in a nursing center, and is a retired journalism and English teacher, also in the public school system. Both enjoy writing and have achieved several awards in contests across America. Gordon served in Vietnam.

Bruce Sydow's poems have been nominated for the Best of the Net and the Pushcart Prize. A combat-decorated marine, he earned his master's degree from the University of Washington in Seattle. A former therapist with Catholic Community Services of Western Washington, Bruce is the recipient of the Sexual Assault Prevention Center's Gold Star Award and the Tacoma Community College Board of Trustees' Faculty Excellence Award. He resides in a seaside village on the Case Inlet of the Olympic Peninsula with the gifted artist, inspiring professor, and constant source of wisdom and love for over a quarter of a century, his wife, Danella Sydow.

N. Jed Todd is a retired Army Master Sergeant, poet, and science-fiction writer. He is an active participant in the Veterans Poetry Project viritual writing workshops. His submissions are published in the *Resilience & amp.*

Catherine Torres is an Air Force airfield operations officer turned writer and librarian. A native New Yorker, Catherine has lived and traveled worldwide. Her work has appeared in regional outlets, *Sapiens*, *EBSCO*, and the forthcoming *Necessary Turbulence.* She currently resides in Bury St. Edmunds, UK with her husband and two children.

Philip Trezza served as an Army and Army Reserve medic from 2004 to 2018, earning the Combat Medical Badge while deployed as a platoon medic in Iraq (2006–2007). Following his military service, Phil dedicated his career to supporting Virginia's veterans and military families. He served as the Director of Veteran Transition & Employment Programs for the Virginia Department of Veterans Services and was instrumental in creating the Military Medics and Corpsmen Program. A contributor to the book *The Fire Within: Shedding Light on Trauma,* Phil continues to advocate for veteran and military spouse employment as a Talent Acquisition Consultant for the Virginia Department of Taxation. He holds a bachelor's degree in political science and an MBA.

Ali Watts is an army veteran with eight years of service and two deployments to Iraq. She resides in Oshkosh, WI and has been published in *As You Were: The Military Review*, Volume 20; *Proud To Be: Writing by American Warriors*, Vol. 13, and *So Long: The Anthology of Poet Warriors*, Volume III.

AFTERWORD
A TESTAMENT OF PERSEVERANCE

Perseverance and resilience. These two words perfectly capture this anthology's mission: to provide a glimpse into the complexities of military service and the lives of veterans. While active-duty members and veterans naturally come to mind when we discuss these traits, committing to this project and reflecting on my own upbringing in a military family has shown me just how rooted and foundational these traits are embedded into the military culture. These qualities are revealed not only through the committal sacrifices of our service members, but also through the invisible, enduring sacrifices of their families and communities. There are those who sign their names on the dotted line to become warriors, and then there are those who were born or married into the lifestyle.

My father was a veteran who struggled with chronic pain before he passed away at just twenty-eight years old. He was an aspiring writer and poet who never had the chance to fully nurture his craft. His memory, along with other members of my family who served, is a driving reason why, after completing my MFA in creative writing, I dedicated this past year to fostering this project's mission and championing veterans' poetry. It was essential to me to encourage veterans to write, to share their work in this anthology, to engage their communities, and to help everyone gain confidence by voicing their experiences through verse.

By volunteering and embracing the challenges of this project, I have gained a profound understanding of what true resilience looks like in practice. Alongside some wonderful partners, Mattie and I have worked to honor the veterans who participated in this civic poetry project and to support their lifelong creative journeys. My hope is that veterans and their loved ones will see reflections of themselves and their experiences—a unique military subculture that often goes unnoticed—and find their own encouragement and endurance.

Megan V. Jenkins
Project and Editorial Team Coordinator

ABOUT THE EDITOR

Mattie Quesenberry Smith, PhD, is an educator, author, and interdisciplinary scholar whose work explores science, technology, engineering, and poetry. A first-generation college student born and raised in the Blue Ridge Mountains, her early life—defined by the rhythms of the New River and the rugged landscape of Appalachia—continues to influence her writing.

Smith holds a PhD in curriculum and instruction from Virginia Tech, where her research focused on describing critical reflective writing within technology and engineering design-based teaching and learning. She is also a Phi Beta Kappa graduate of Hollins University, holding a double major in biology and English literature, as well as an MA in English literature and creative writing.

For over a decade, Smith has served as a faculty member at the Virginia Military Institute, teaching rhetoric and writing. Her teaching career includes service to the Virginia community college system, where she mentored students across diverse disciplines. A dedicated community facilitator, she has long championed local voices, organizing creative writing workshops and community reading series such as Writers at Studio Eleven to foster connection among professional, academic, and local writers.

As a writer, Smith shifts fluidly between genres. Her poetry and prose have appeared in numerous literary magazines, including *Southern Poetry Review, Dappled Things*, and *New Verse Review*. Her recent writing can be found on *poets.org*. Her chapbook, *Mother Chaos: Under Electric Light*, was nominated for a Library of Virginia Literary Award. Her professional portfolio includes experience as a legislative assistant and ghostwriter for a US Congressman, and a book reviewer for the *Hollins Critic*.

Beyond the page, Smith is an award-winning screenwriter and documentary filmmaker. Alongside her husband, Douglas N. Smith, she has garnered several film awards, including a CINE Eagle award and first place in the New York International Film and Video Festival for their documentary *Between Two Fires*. Currently, the Smiths are developing a multi-platform project adapted from Wallace Breem's *Eagle in the Snow* which explores a catastrophic shift in the course of Roman history when the Rhine River froze on New Year's Eve, 406 A.D., allowing 40,000 barbarians to cross into Germania.

Smith resides at the foot of Little House Mountain in Lexington, VA, where she and her husband are visited often by their ten children, extended family, and a host of dogs.

Please visit her website at https://mattiequesenberrysmith.com.

PREVIOUS PUBLICATIONS

Faithworks Publishing first published David Reid Brown's poems "Global," "Deployment," and "Home."

Kenneth Conklin's poems "Agent Orange!" and "Post Trauma" are previously published in *The Zen of Ken* in 2021.

A shorter version of Jeffrey Parker's "One Soldier; One Freedom" appeared in the December issue of *KAKALAK* 2025.

Laura Wence Connor's poem "Re-integration" was originally published in *Proud to Be* Volume 12, and her poem "Quiet Moments" was originally published in *Proud to Be* Volume 13.

A SPECIAL THANKS

I am thankful to all our partners and the wonderful teams we have collaborated with in launching the Veterans Poetry Project and publishing this anthology.

I want to give special thanks to Nadra Mabrouk and Mary Sutton at the Academy of American Poets for their unconditional encouragement.

I also want to thank Dr. Kathleen P. Decker and Wayne David Hubbard for their faithfulness throughout the design and enactment of the Veterans Poetry Project.

I am grateful for Jeanne Johansen from High Tide Publications for her patience and knowledge of the publishing industry in helping to bring this book to life.

Finally, I want to thank Megan V. Jenkins for the months of behind-the-scenes volunteer work she has given unconditionally to facilitate this project and its anthology.

My family deserves special thanks for their unwavering support for this work—I could not have realized this dream project without my husband, Douglas N. Smith, our wonderful children, and all their spouses and partners!

- Mattie Quesenberry Smith

www.ingramcontent.com/pod-product-compliance
Lightning Source LLC
Chambersburg PA
CBHW041302210926
51847CB00010B/116
* 9 7 8 1 9 6 2 9 3 5 7 3 9 *